High Tech Among the Palmettos

The Story of Radiation, Inc. and How It Changed the Face of South Brevard County

by Frank Perkins

and

a bunch of old Radiators

© 2014 Frank Perkins

All rights reserved. No portion of this book may be reproduced or used in form or by any means without the written permission of the publisher.

Manufactured in the United States of America

Oak Publishing

5225 Crane Road

Melbourne, FL 32904

Contents

Chapter 1	Historical Summary	13
Chapter 2	The People	45
Chapter 3	Personal Memories	65
Chapter 4	Plants and Facilities	145
Chapter 5	Major Programs, Products and Systems	153
Chapter 6	Other Programs	179
Chapter 7	Radiation Printer	197
Chapter 8	Stonehouse Tales	203
Chapter 9	Branches, Acquisitions and Spin Offs	217
Chapter 10	Memorable Customers	221
Chapter 11	Timeline	225
Chapter 12	The Community	231
Chapter 13	Publications	243
Chapter 14	The Ongoing Spirit	253
Chapter 15	Musings	257
Chapter 16	Where to Find More Material	261

FOREWORD

"THERE'S NOTHING BUT ALLIGATORS AND MOSQUITOES DOWN THERE! FORGET ABOUT IT!"

George Shaw says that's the response Homer Denius got when he recommended to his boss at Melpar in Virginia that a division of the company be established in Florida. George and Homer had just witnessed the first successful launch of a V-2 rocket from Cape Canaveral in 1949, and they both felt strongly that there was good future business potential in space electronics so they sold their homes in 1950 and used the sales proceeds to establish a new company they named "Radiation, Inc." in a surplus Navy building at the Melbourne airport. The decision to set up operations in Melbourne wasn't hard – George had some slight familiarity with the area from earlier visits while attending the University of Florida, the rent at the airport was cheap, and perhaps most important of all it was a great place for boating!

The building leaked, the only bank in town wouldn't loan them money against Government contracts, and yes, there really were mosquitoes by the millions, but they persevered. Their first employees were an accountant hired locally and a few engineering graduates from the University of Florida, and George recalls that local folks thought they were all electronic wizards when they erected an antenna on top of their rented two-story building and succeeded (sometimes) in receiving fuzzy television images from Jacksonville.

During the first year or two Radiation received contracts to design and install an interim timing system along the Eastern Test Range interconnecting the various optical and telemetry and radar instrumentation, and a contract to supply a team of engineers to help the Glenn L. Martin Company design and deploy a new guidance system for their sub-sonic, air-breathing Matador missile – one of the first users of the new test range. And through contacts with folks at Wright Air Development Center that Shaw and Denius had made earlier while they were still at Melpar, they also received a few other contracts related to flight-testing of aircraft. Test aircraft were typically instrumented with devices for measuring technical parameters like temperatures and forces and vibrations being experienced by the airframe, and these measurements needed to be gotten back to the ground for human interpretation. One option was recording the parameters onboard the aircraft for later readout, and the better option was to telemeter these parameters – that is to send the measurements back in real time via radio links. Radiation received contracts to do both, but telemetering turned out to be the company's forte, and the key to its ultimate success and growth.

By early 1953, with its headcount approaching 100, most administrative and service functions as well as some engineering functions were moved to a second rented building located about where Melbourne's airport terminal building is now. The original building stood a few blocks south, adjacent to the Trailer Haven Recreation Hall. Many things have changed in the ensuing sixty plus years, but Trailer Haven is still there, little changed from the 1950s. Radiation's most important project underway in the early 1950s was a contract from WADC to upgrade the AKT-6/UKR-1 radio telemetry system originally designed by Melpar, and then to go on and manufacture several for use in testing the susceptibility of jet aircraft to atom bomb blast effects during the 1955 Teapot atomic weapons test series in Nevada. But of course that intended use was all a big secret in those days, and when Radiation was awarded a follow-on contract to actually instrument several QF-80 drone jet aircraft with the telemetry equipment for the tests, the company wasn't able to tell the Melbourne City Commissioners why the airport's runway needed extending to the 10,000 feet the jets would require for takeoff. Frustrated with the Commission's unwillingness to extend the runway, Radiation established a new Instrumentation Division in Orlando, adjacent to the long runways at Pinecastle AFB (now McCoy).

The success of that contract led the Telemetry Lab at WADC to award Radiation an additional contract to pick up the pieces from another only-partly-successful effort by Melpar to develop the AKT-14/UKR-7 PCM telemetry system, and this represented a real turning point for the company – the first practical application of Pulse Code Modulation telemetry. The Cold War was going full blast by this time, and our end of the missile race with the USSR had graduated from the likes of the Matador and Snark air-breathing missiles to ICBMs and IRBMs like the Atlas and Thor, which demanded far more robust and rugged telemetry systems than the elementary FM/FM equipments that had represented the Inter Range Instrumentation Group's standard in missile testing up to that point. With this higher capacity AKT-14/UKR-7 as a starting point, and with a determined management and a sometimes unscrupulous sales manager, Radiation was able to "corner the market" for PCM telemetry systems by under-bidding and over-promising to such an extent that for awhile it found itself late and overrun on most jobs; but fortunately one more large and very high priority contract was booked that finished paying for the development, so as they say, "all's well that ends well!"

Finding itself at the forefront of radio telemetry technology for Cold War missile testing and space race applications, Radiation basically organized itself to accommodate three related but disparate specialties. First there was Aerospace Division, whose main focus was on miniaturizing and packaging of the space-borne PCM telemetry elements to withstand the harsh environments associated with rocketry and space. Second there was RF Division whose main focus was on developing high-gain antennas for tracking the rockets and spacecraft and recovering the weak radio telemetry signals from great distances. Third there was Ground Data Division whose focus was on computer processing and interpretation of the recovered PCM digital data. And overlaying all of these was an Advanced Technology function, whose focus was on techniques for trying to squeeze the maximum theoretical performance from the power-limited links characterizing radio telemetry -- power-limited because telemetry was usually something of an afterthought, hitchhiking a ride and competing for power with payload apparatus. Thus, a mindset and culture of "efficient communications" evolved within the company. As the Cold War wound down, the expertise gained in these telemetry specialties carried over gracefully into peacetime endeavors like space exploration, space communications, satellite meteorology, satellite reconnaissance, etc., positioning Radiation, and now Harris Government Communications Division, as the "go-to place" for many of these, and related, specialties for both commercial and Government use.

By the time of its merger with Harris Intertype in 1967, Radiation had more people on its payroll than the entire population of South Brevard in 1950. And these were extraordinary people, drawn from top universities and companies by the opportunity to do challenging work in an engineer-managed company, and to bask in Florida's sunshine. Realizing early on that in order to attract and retain extraordinary people there needed to be good schools, good hospitals, good shopping, good banking, affordable housing, and the like in the community, Radiation worked with civic leaders and others to bring about change in the once-sleepy villages of South Brevard. When Florida Institute of Technology was little more than a gleam in Jerry Keuper's eye, Radiation provided direct monetary support, its staff taught night classes at the college, it provided tuition grants so Radiation employees could attend the college, and through political contacts it helped the college pursue grants and the like. George Shaw was a founding trustee, served a term as an early Chairman of the Board of Trustees, and continued active on the FIT board for about 50 years. George also worked tirelessly to promote unification of the several separate communities for more "clout" in Tallahassee and Washington, and he worked with Jimmy Holmes in spearheading efforts that led to replacing the inadequate Melbourne Hospital with the modern Holmes Regional Hospital of today.

Denius and Shaw are both gone now, and the name "Radiation, Inc." is largely unknown to present-day residents of South Brevard, but the shrinking number of folks who worked there still remember, and still gather annually to retell stories of the "glory days" fifty and sixty years ago when great things were happening and when even overtime work was fun. I'm glad to have been a part of those times!

The foregoing is simply one old timer's perspective on the high points of how Harris Corporation and South Brevard County got to where we are now. Subsequent chapters of Frank's book will provide different perspectives and fill in lots more detail.

A.B. Amis – August 2014

Preface

A group of retired employees of Radiation, Inc. gather annually to refresh old memories. A consistent theme is what an enjoyable place it was to work. And not because it was easy. On the contrary, the hours were long and the challenges were extreme. Now we can recognize how close to the cutting edge of technology we were working, with scant experience and austere resources– and all because Homer and George had promised the customer that we would do a good job.

As examples of early day challenges and achievements were recalled it became clear that many were worthy of preservation. Copies of pictures and documents were made and exchanged, but most still resided in storage boxes under the spare bed. A wiki-style web site was created to organize and preserve some of the memories. Preservation of this data in either electronic or original physical form for the next generations is difficult to assure in an accessible and reliable manner.

A printed book offers advantages. It can be organized in a coherent manner, and can be enjoyable to a variety of people, including those not comfortable with navigating the computer world. It can find long term homes on library shelves and home bookcases.

Thus, the purpose of the book is to preserve the history of Radiation, Inc in a durable manner and in a form that is interesting to former Radiators, their friends and descendents, future researchers and others interested in technology, history, people, places, organizations and relationships among all these things.

Enjoy!

Acknowledgements

I would like to express my appreciation to all those who shared there memories and documents. When I began collecting these I did not envision a formal archive, so I did not record exactly who contributed what. Thus I am unable to to provide accurate, complete credits, but that does not alter my appreciation.

Chapter 1

Historical Summary

This book is a record of the goings-on at Radiation, Inc., in the 1950's and 1960's, plus a few stories from outside this period. It is intended to provide a record of the history of the company and the community, and a summary of the recollections and memorabilia of some of the employees of that period.

Radiation was founded in 1950 by Homer Denius and George Shaw. At the time they worked at Melpar, Inc., in northern Virginia. Homer and George presented plans to the Melpar management for establishing a telemetry division in central Florida. Melpar rejected this proposal, so Homer and George decided to try it on their own. They chose Melbourne, Florida as a site, ostensibly because of its proximity to Cape Canaveral and its missile range activities. Actually, the good boating in the area was probably a major factor to the founders, since they were both enthusiastic yachtsmen. Melbourne also offered the virtue of inexpensive space in buildings at the Melbourne airport, dating to its World War II service as a Naval Air Station.

The company was funded from the savings and the borrowings of the founders, which was limited, but gave them the freedom to run the company and pursue opportunities as they saw fit. The name confused some people, who associated it with atomic radiation; actually it referred to radio frequency radiation, i.e. communication by radio.

In the early days, Radiation took any interesting business they could get, often marketed primarily on the basis of a best effort on a challenging problem.. This approach was abetted by a talented staff, recruited from the top engineering schools. These people were attracted to Radiation by the promise of an engineer-led organization, working on challenging programs. The Florida location didn't hurt either. They played hard on the water and beach, but worked harder, and for longer hours. Homer is reputed to have ended a marketing presentation with "We have some very good people for this job, and if you give us a chance, I know we'll do you a good job!"

The company soon developed expertise in several areas, including: PCM telemetry systems (airborne and ground); digital data acquisition and processing systems and large tracking antenna. In many cases they were working decades ahead of the practical state of the art. They struggled with imperfect components to accomplish functions requiring many racks of equipment, functions which are now performed in everyone's smart phone.

During this period, most of our customers were very patient and understanding. Bill Eddins is reported to have said "We build equipment that doesn't work for people who don't need it!" Actually, I believe that most of the customers were purposely pushing the envelope and were more interested in finding out what could be done. As long as they knew we were trying hard, they were satisfied.

The first customer that was different was Boeing. They were very serious about every specification and contract requirement, significant or not. Once we understood this, it became a challenge that we eventually overcame., and we were a better organization for it.

During the 1950's Radiation built or acquired facilities in several places across the country. In Orlando, a Research Division and a manufacturing facility were established to take advantage of the larger labor pool and larger infrastructure there. An Instrumentation Division was also established, originally to install instrumentation in aircraft which needed the longer runways in Orlando.

Radiation also felt the need for a presence in California to be near the air and space activity there. In 1959 they purchased Levinthal Electronic Products and established Space Communications Division, both in the Bay Area. The Levinthal acquisition supposedly came about because Homer and Elliott Levinthal were seated together on an airplane flight. Levinthal specialized in high power transmitters and related power supplies and modulators. They built equipment for the Arecibo Radar in Puerto Rico, and the DEW Line radars.

In 1959, Radiation broke ground on a new facility in Palm Bay, and by the end of the year were occupying the first three buildings of what eventually became a large complex there.

In 1967, Radiation was merged with Harris-Intertype, a printing equipment manufacturer in Cleveland, Ohio. Radiation wanted the merger because it was felt that integrated circuit manufacturing capabilities were needed to remain competitive, but that the investment needed would stress the company. Harris-Intertype wanted the merger because they wanted to branch out into a broad communications company. The merger was eventually reasonably successful, although the organizations had quite different cultures, but clearly an era was ending, and we conclude our story here.

An example of the vision in the company in those days is the following George Shaw memo, of September 1959.

This is followed by a reprodction of a "Welcome to Radiation brochure of the 1950's.

MEMO

DATE: 11 September 1959

TO: L. P. Clark
Dr. Chas. R. Burrows
Dr. J. Q. Brantley

cc: L. W. Sieck

FROM: G. S. Shaw

SUBJECT: Future of Space Programs

National survival now requires that we immediately attain an invulnerable offensive striking force. "Polaris" is such a force; although it has operational communications, manpower, and maintenance problems. "Minuteman" will be hardened to provide partial invulnerability.

A stable world military pattern will result when both USA and USSR have the invulnerable offensive force. By stable, I mean that war cannot accidentally start. It could only start intentionally, after one nation thought it had achieved the enormously expensive "perfect defense".

It is conceivable that radar-small armed satellites with precision re-entry and targeting computation capability may some day be added to these invulnerable offensive forces. The task of locating the satellite, which is by nature methodical and predictable, appears easier than locating a randomly traveling submarine.

It is probable, that as the world enters an era of the invulnerable offense (and such an era can exist for a number of years) there will be a concentration of national efforts on the minds of uncommitted men.

Communism controls 1/4 of the earth's land area and 1/3 of the earth's people.

Democratic capitalism does not in the same sense control any of the earth. Democratic capitalism is North America and Western Europe. It is 1/4 of the land area and 1/6 of the earth's people. 1/2 the world's land and 1/2 the world's people are uncommitted.

The world communication satellite is the ideal way to reach the uncommitted half of the world. It is one of the immediate, truly valuable, down-to-earth reasons for a
Space program. (The reconnaissance satellite is another valid reason).

The man-in-space, the lunar, and the planetary probes are basic research programs since we cannot now conceive of their direct value to the nation and its people. The logical reason for such research programs is to increase the world's fundamental knowledge. The speed with which we are now trying to accomplish this basic research does imply that we are more interested in impressing the world with "firsts" than providing fundamental knowledge.

World-wide communications via the earth's ionosphere, radio circuits, and submarine cable is a scant 10 megacycles of utilizable bandwidth on a 24-hour basis. Less than two television channels of bandwidth now convey the world's commerce and news.

A hovering three-satellite system utilizing existing technology can duplicate present world-wide communications capability. Expansion of satellite equipment content and the addition of new systems of satellites for world communication can provide 1000 to 1 increase in capability within a decade. World commerce, news, education, and entertainment can also expand 1000 to 1.

The USSR will enter a world communication program based upon satellites. It will be on a nationalized basis – just as is their Aeroflot, telecommunications, and power generation. They will do this in their quest for the uncommitted world minds.

The USA government must provide the equatorial launching site (Christmas or Galapagos Islands) and the launching hardware. Our government must also provide the sample system, suitable for military and governmental world communications. The government could also provide satellite relay bandwidth for commercial, education, and entertainment circuits. However, commercial interests would provide the follow-on program if they could purchase government launch and injection capability. Commercial interest is assured, once technical feasibility is demonstrated, because a communication unit (such as a kilocycle-mile-minute) can be furnished for between 10^{-3} and 10^{-5} present land-based communications costs.

With three years of study and preliminary development this company could be prepared to provide the initial commercial satellite relay in the 4th or 5th year, for a total of 15 to 20 million dollars. Joint support is quite possible, from world oil and mining interests, a broadcast network, a telephone firm, and a telegraph firm, each of which do not now have the nucleus of the required technical capability. In fact, a 20 million dollar support program might be achieved by Radiation from the several mentioned sources, more quickly than it can be acquired from the individual Boards of AT&T, ITT, or RCA.

Toward this end, Radiation should endeavor to play as active a part as possible in programs leading to a hovering communications relay satellite. We should endeavor to find suitable space communications programs early, at either BMD, WADC, or other sources. There is indication that a team of smaller companies is desired for some major space program for which they can demonstrate competence. Small Business Administration has such current dollars and will assist.

Our present studies and participation in project Score and Courier help demonstrate competence in certain areas. We should strive for a small portion of the GE-Bendix Notus project.

We also need an "on-board" program as well as the present ground support type programs. Our telemetry experience may be the best to get us on board the vehicle.

G. S. Shaw

[The world's first active communications satellite, Courier was launched October in 1960, and used a Radiation tracking antenna. Telstar, the first communications satellite to relay TV signals across the Atlantic was launched in July, 1962 carrying a Radiation PCM System. The electronics for Hexagon, the largest reconnaissance satellite, launched in July 1971, was designed and built by Radiation.]

A Brochure from the 1950's

WILLIAM E. KEPNER

HOMER R. DENIUS

GEORGE S. SHAW

WILLIAM W. DODGSON, JR.

INTRODUCTION

RADIATION, INC is an engineering firm engaged in research and development in electronics, avionics, and instrumentation. Office, laboratory, and field facilities are located in seven plants at three airports; main offices and engineering facilities at the municipal airport at Melbourne; production and instrumentation facilities at the Orlando airport; and radar testing and measurement facilities at a leased auxiliary airport near Melbourne. Communications are maintained between all plants by leased telephone lines, leased and company-owned-aircraft, and company cars.

OFFICERS

WILLIAM E. KEPNER	Chairman of the Board
HOMER R. DENIUS	President
GEORGE S. SHAW	Vice President – Engineering
W. W. DODGSON, JR.	Vice President – Contracts
GRACE DENIUS	Secretary
JOHN W. BOONE	Treasurer
RANDOLPH MATHENY	Legal Counsel
WILLIAM E. KERCHER	Controller
RALPH L. BICKFORD	Director – Industrial Relations
FRED A. CULLMAN	Director – Production Div.
ROBERT DRYDEN	Director – Instrumentation Div.

ORGANIZATION

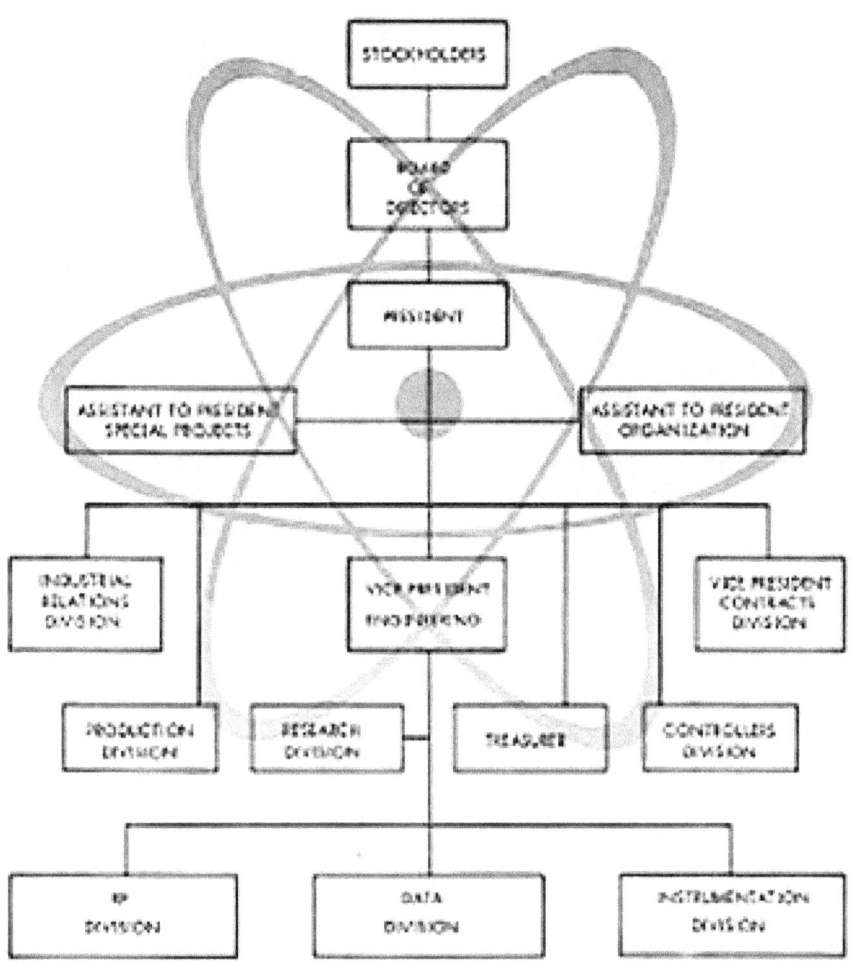

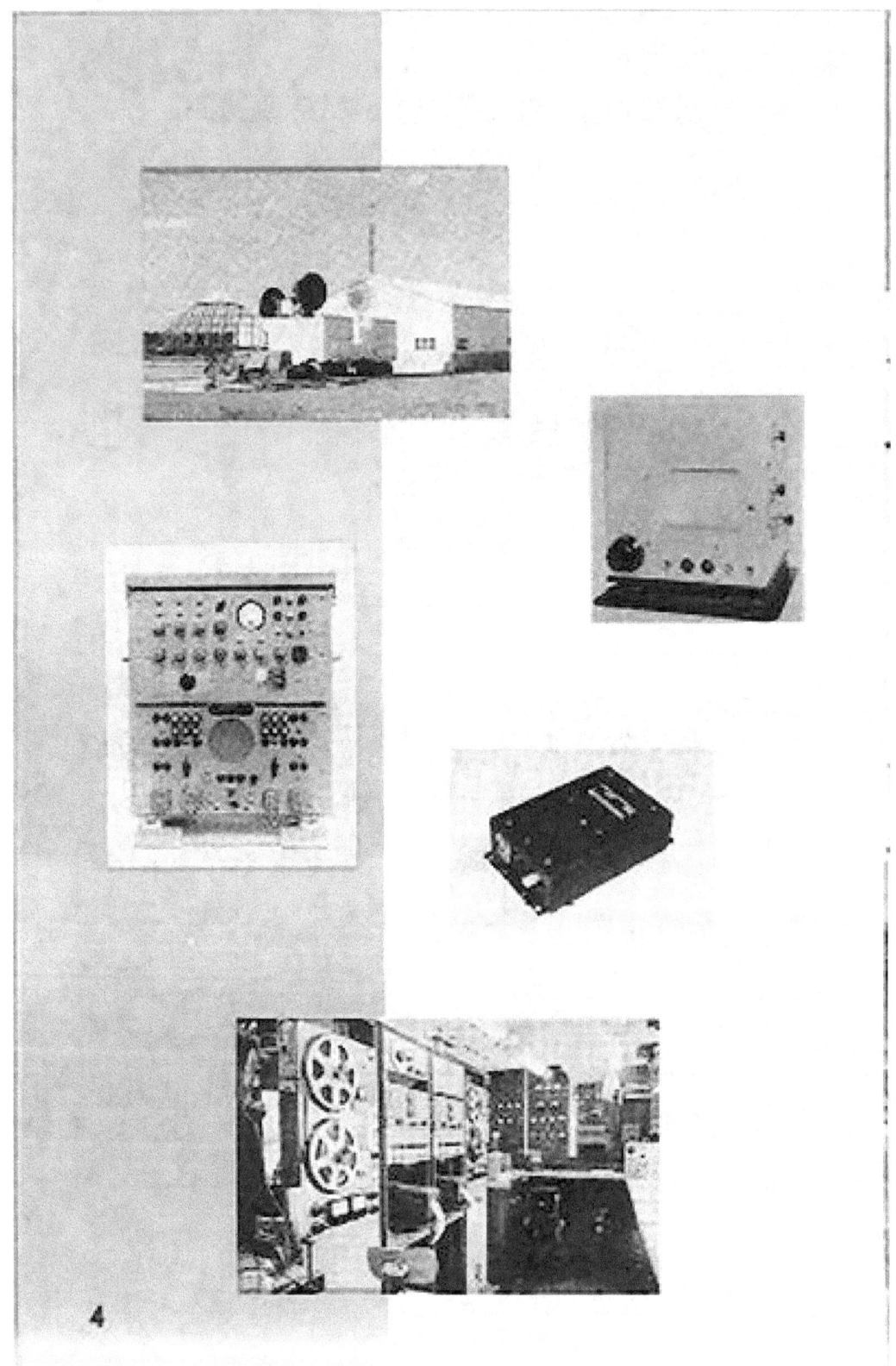

4

ENGINEERING

Work of the engineering division is carried on under the supervision of the various project engineers who report directly to the respective section heads. Project engineers are in charge of all phases of a project --- administrative as well as engineering --- and delegate responsibility for the various phases to assistant project engineers.

The engineering division is organized into three main sections: design, development, and instrumentation. Most of the work done by RADIATION falls into one of these primary categories. These classifications are necessarily broad and may involve anything from a 60-foot antenna to a subminiature transmitter, or from research into the characteristics of a transistor to the complete instrumentation of a guided missile.

PLANT 1

ENGINEERING AND SERVICES

Plant 1 at Melbourne contains engineering laboratories and offices of the test equipment group, accounting, personnel, and purchasing functions, and storage for records and materials.

ENGINEERING

Plant 2, across the street from the Melbourne airport terminal, is completely modernized and air-conditioned and serves as the "main" plant of the company. Here are located all administrative functions as well as special purpose laboratories, engineering offices, drafting services and a complete machine shop. A "special processes" lab which provides chemical and painting facilities is adjacent.

PLANT 2 AND ADMINISTRATION

PLANT 3 PRODUCTION

All facilities for the mass production, assembly testing and inspection of electronic equipment are contained in the two buildings of Plant 3 at the Orlando airport.

PLANT LOCATIONS

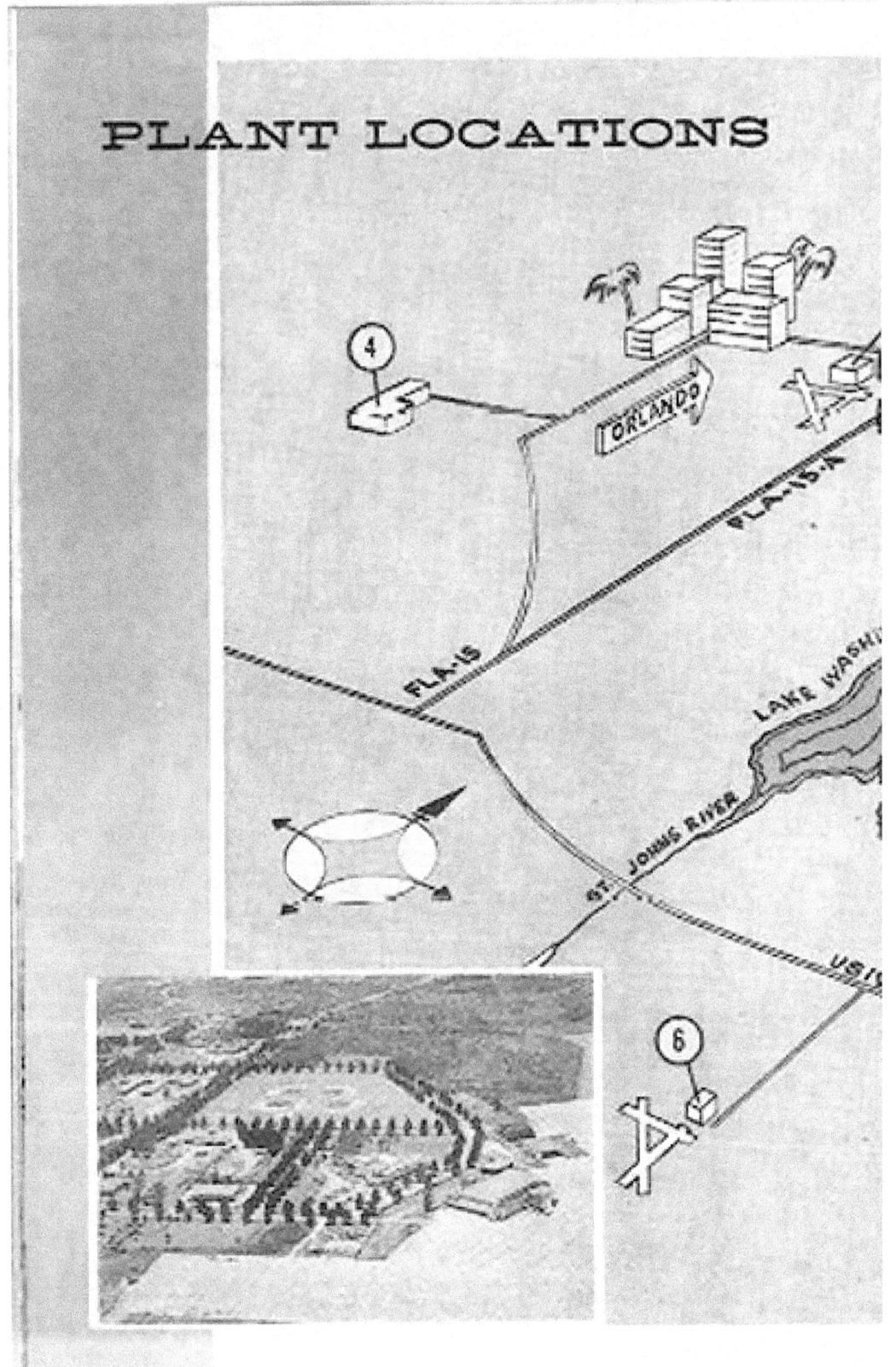

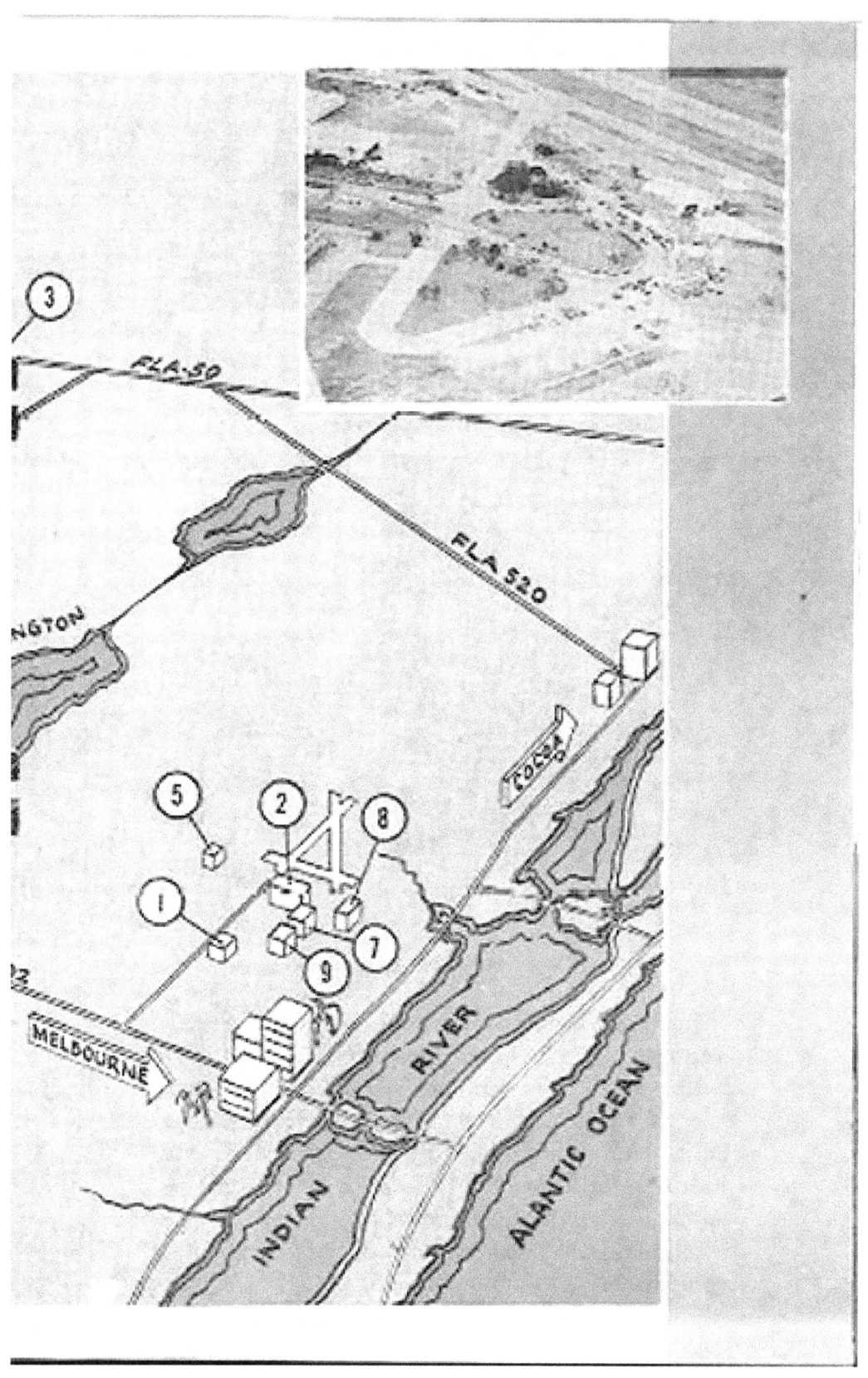

PLANT 4

Hangar and engineering spaces for the Instrumentation Division are provided by this new building adjacent to Pine Castle Air Force Base near Orlando. Here telemetry trailers are built and equipped, aircraft modified and instrumented, and procedures and designs set up for the installation and testing of transducers, recorders, cameras, and other instrumentation equipment.

INSTRUMENTATION

PLANT 5   RADAR

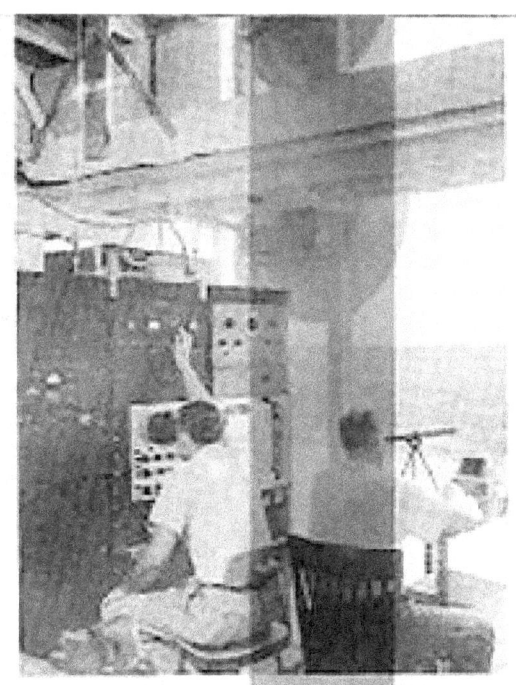

MEASUREMENT

This building houses the facility for making radar reflectivity measurements, and recording and reducing the obtained data. Work is carried on here under contract to both governmental and private agencies. The location of this site is near the Melbourne airport.

17

PLANT 6

This facility is located at a leased auxiliary airfield near Melbourne. It contains complete equipment for the testing and calibration of land and airborne radar systems.

RADAR TEST

PLANT 7

This building provides laboratory and office spaces for the digital section. It also contains a completely equipped test area for the type testing of a wide variety of electronic units. These facilities, the most complete in the Southeast, include altitude, humidity, shock, vibration, temperature, and pressure environmental test equipment.

DIGITAL EQUIPMENT

PLANT 8

PUBLICATIONS

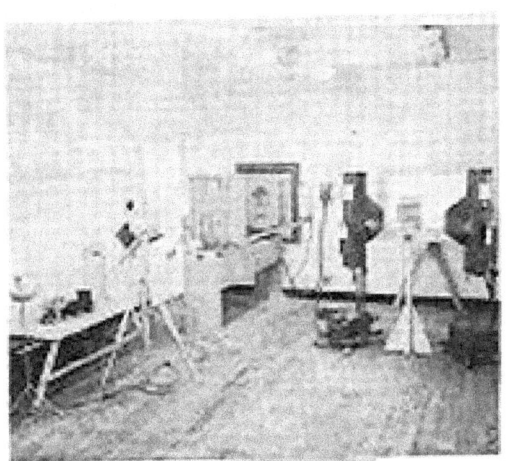

Located adjacent to plant 2, this building contains photographic and printing facilities of the company. Here the publication section produces reports, instruction books, and internal forms and publications.

PLANT 9 ANTENNA

This building, at the Melbourne airport, houses labs and offices of the antenna development section. This section is involved in a comprehensive program of design, modification, and evaluation of various radar and telemetric antennas and systems.

DEVELOPMENT

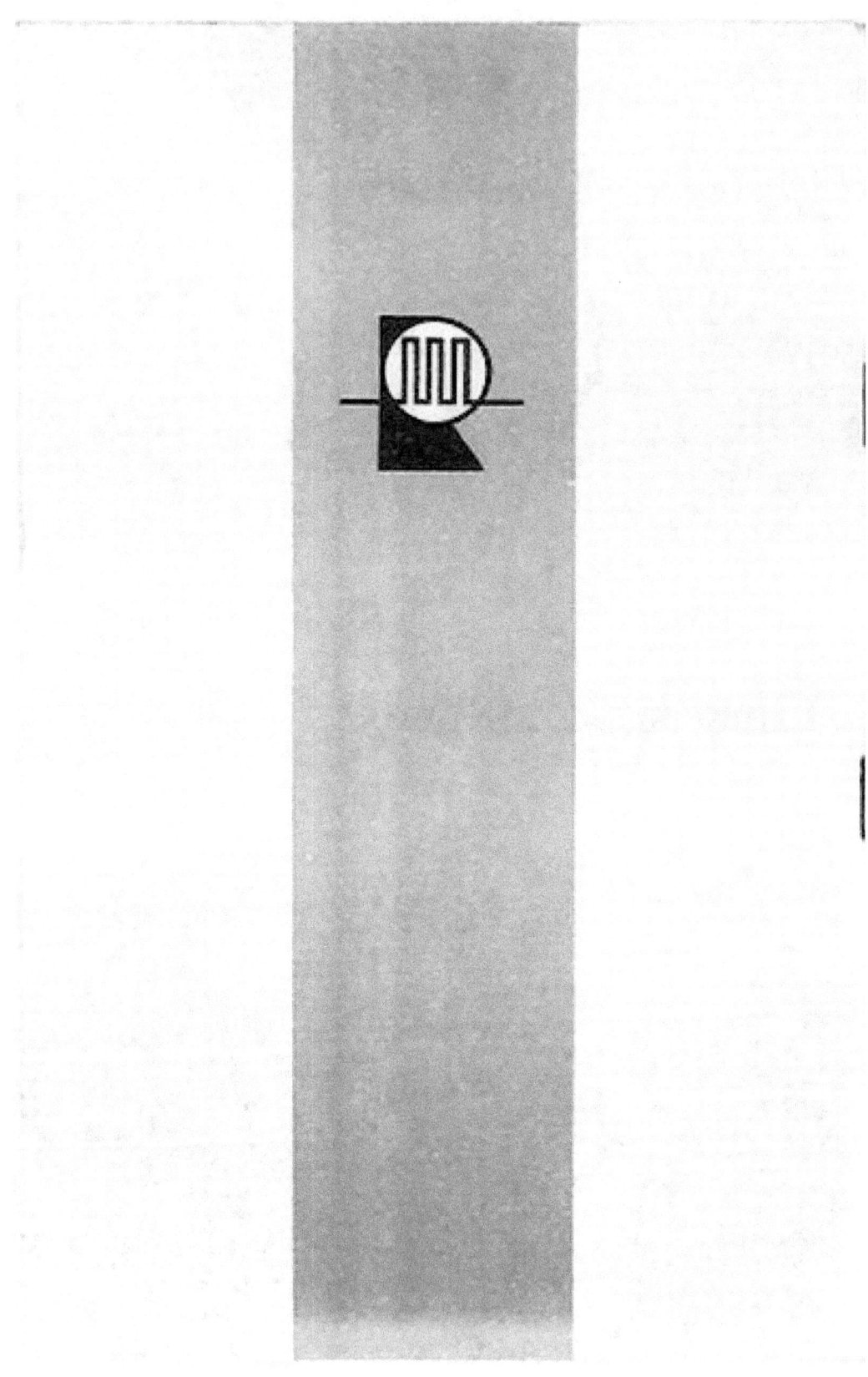

Chapter 2
The People

The strength of Radiation was a reflection of the dedication and perseverance of many of the very early employees. To discuss these people I choose as my guide the earliest company brochure I possess, from about 1955 (with a stylized guided missile on the cover), because it has an extensive list of early employees.

The "key employees" listed in this brochure include co-founders Homer Denius and George Shaw, as well as William Dodgson and John Boone.

Homer Denius

Homer was born January 31, 1914 in Hamilton, Ohio, and started early in electronics. He reports, "When I was in the fifth grade in southern Ohio I made my first radio set out of a cardboard oatmeal box and a crystal. Somebody gave me a set of headphones and some wire and I could get KDKA in Pittsburgh. Then another fellow came along and gave me a lot of radio parts — a three-tube set and a whole bunch of parts to make radios out of. That's what got me started, and I never got away from it."

> *[Editor's note: Oatmeal is still packaged in a 4" diameter, 7" long container which is the ideal size for winding the coil for a broadcast band receiver. A variable air capacitor tunes the coil to the appropriate frequency. The detector for my receiver was a do-it-yourself semiconductor diode. A piece of Galena was contacted by a thin wire "cats-whisker". Over time the crystal would oxidize and fail. The solution was to gently scape the surface of the crystal until it was again shiny, and to experimentally find a "sensitive spot" with the cats-whisker. Voila, the receiver was as good as new. A good pair of earphones was the only other part you needed. A 50,000 watt transmitter would provide audio you could detect across the room. My set had a hemispherical glass cover for the crystal that looked really cool, but the crystal still oxidized in a day or so. I do not understand why this configuration produced a rectifier, but early transistors used a "point contact" structure that must have been similar. Wikipedia has an informative article on the cats-whisker.]*

Homer received his education at the University of Cincinnati, majoring in Radio Engineering. Prior to founding Radiation, he worked at a number of companies including Crosley Corporation, Magnavox Company, Ken-Rad Tube Company, American Type Founders Corporation, and finally Melpar in 1946, where he was Vice President, Director of Engineering and member of the Board of Directors. He was a co-founder of Radiation, Inc., and served as President and contributed greatly to marketing operations.

Of the early days of Radiation Homer reports: "The fact that we were able to get in the digital space technology business before anyone else, and did a very good job in that respect, was largely responsible for the growth of Radiation. We didn't have much competition and our group had more knowledge from the start in digital telemetry than any other group in the country. I don't think we ever lost a contract in that particular area."

"One big lesson I learned when I was quite young, and I've always followed it. Never make a promise you can't keep. And it's really paid off. It doesn't matter, as far as I'm concerned, whether it's business, or friendship, or something else. You should never, never make a promise if you can't keep it. That's the way I feel about things."

Homer was a force in the yachting world. Sports Illustrated magazine of March 22, 1965, reported on his ocean racing yacht Maredea: *"When Space Scientist Dr. [sic] Homer Denius of Melbourne, Fla., a man who likes to be No. 1, asked Charley Morgan to build him a new boat, he specifically asked for a craft that would be 1) the most comfortable cruising boat possible, and 2) the fastest Class-A ocean racer in the world. With her 14-foot beam, 40-foot waterline and 1,550 square feet of working sail, Maredea—the biggest fiberglass boat of her kind ever built—fulfilled the first requirement easily. She has one more cabin than any other boat her size, 2 tons of air conditioning, hot and cold running showers, complete cabin ventilation system and a deep freeze that, even open, keeps butter so hard one has to cut it with a hacksaw. The second part of Denius' request was more difficult."*

But she did win races.

Sports Illustrated of July 08, 1968 describes another of Homer's yachts:

"One way [to win race] is to build a boat so swift that no other craft of her class can sail with her. One such this year was Rage, a magnificent 53-foot sloop designed, built and sailed by Charlie Morgan of St. Petersburg, Fla. for Homer Denius. Constructed more like a miniature 12-meter than an ocean racer, Rage piles so much sail on her tall spar and carries it so well that even with a higher rating she has handily beaten most other boats of her size. She did it again last week when she took Class B honors and second overall." This was in the 1968 Nassau Cup race.

Someone once asked Homer how many boats he had owned. He responded: "Over 30 feet? About 60."

Homer was a big supporter of Florida Institute of Technology. The Denius Student Center on the campus is home of The Office of Residence Life and Office of Student Activities and is the hub for all student clubs and organizations.

Homer passed away on April 20, 2006 of complications of Alzheimer's disease at the age of 92.

George Shaw
George S. Shaw, Vice President and Director of Engineering, was born in 1921, in Brattleboro, Vermont. He received his B.E.E. Degree from the University of Florida and did graduate work at Columbia University, majoring in UHF techniques.

He engaged in klystron research at the Sperry Gyroscope Company on Long Island and later was assigned to the Naval Research Laboratory to assist with the development of multichannel airborne transceivers. George served as Microwave Consultant to the International Business Machines Company and later became a Research Engineer on proximity fuses at the University of Florida. Mr. Shaw joined Melpar, Inc. in 1948 and held the position of Project Engineer, His major endeavors there were devoted to the design and development of telemetering systems and various instrumentation and microwave developments. He is a co-holder of patents and patent applications on electronic circuit miniaturization techniques, a high-altitude stabilized balloon platform for radar and television transmissions, and digital data handling equipments. He was experienced in micro-wave design, and made contributions to developments in the fields of pulse time and pulse code communication, sub-miniaturization techniques, data recording and processing equipments, instrumentation of aircraft, and pulse and micro-wave test equipment. George was a Registered Professional Engineer in the State of Florida, and a Senior Member of the Institute of Radio Engineers.

In 1950, George and his friend Homer Denius started Radiation in Melbourne, "whose major industries were dairy farmers and mullet fishermen." George visualized that the world would soon change as the space age approached. He spoke about the future at local meetings, on radio and, later, television shows. In the 1950's, when he predicted satellite antennas on the rooftops of houses around the world receiving hundreds of broadcasts, many thought he was talking very strangely.

Larry King interviewed him on his radio show. George said, "He thought I was a wishful thinker."

When asked about the challenges to get his visions accomplished, Shaw said, "The biggest problem I had was getting people. We were engineer-limited in the beginning, and the early development of the company was dependent on getting good people. We could see right away that something had to be done for the community in order to make it attractive to bright, young people coming out of MIT, Georgia Tech, or Purdue. How in the world do you get them to move to a town of 5,000? How do we get people to visit and find out what a lovely place Melbourne would be to live?"

Shaw became actively involved in the community to solve this problem.

"Very early on, the need was seen for a school here, and so the Brevard Engineering College, later named Florida Institute of Technology, was started by Dr. Jerome Keuper and of course Radiation supported it 100 percent," said George.

Another problem was the hospital. "The hospital was kind of like a 20-room motel down on Route One. I served on the hospital board with Jimmy Holmes, and the Holmes Regional Medical Center now exists because of our work and our pushing in that direction."

The community had a need for a new bank. Shaw remembered, "The one here was a pretty poor bank. So Mr. Denius and I, along with a group of local businessmen started the First National Bank of Melbourne. The bank was responsible for starting the Trinity Tower development here for the elderly, and it also helped finance Florida Tech. The First National Bank of Melbourne was sold to the Sun Group after the merger of Radiation and Harris-Intertype."

Shaw added, "We built the community from the inside. I was commodore of the Melbourne Yacht Club one year and two years at Eau Gallie Yacht Club. The clubs gave people something to do and a place to go which helped attract the new engineers."

With all the positive things happening Shaw said, "There was one crucial point of conflict with the community. We had acquired a contract to digitally instrument some aircraft to fly through nuclear tests. It was an extremely important contract for us. The planes could be trucked in, but after we did our work, they had to be flown out intact."

"The runways at the Melbourne Airport were too short. We offered to pay for extending the runways in exchange for use of some vacant World War II buildings we needed, but the city wanted us to pay rent too. One of the difficulties was that it was impossible to talk about what we were doing. The project was highly classified. All I could say was that we have to have the runway longer. I couldn't tell them why."

"Fort Lauderdale, Orlando, and Miami were willing to do almost anything to get us to move there. We wound up building our own facility for the project in Orlando, where the runways were long enough, but we almost left the Melbourne area."

"At that point, we could see that if we were to keep up with the business we were getting we had to get into some substantial and adequate facilities. Neither Homer nor I wanted to move to Orlando. The people in Palm Bay made us welcome and helped us find financing, and that's how we came to locate our main facilities there in 1960."

After the merger with Harris-Intertype, Shaw stayed until the end of 1968 when he and the other three top Radiation executives, Homer Denius, George "Art" Herbert, and later Johnny Boone resigned to devote full time to a new company they started called Electro-Science Management based in Ft. Lauderdale. Because it did not require much time from Shaw, he went looking for new adventures. Soon he purchased a large parcel of land in Costa Rica and along with his sons, Dale and Alan Shaw, they helped create a new small town. The sawmill, cattle ranch and coconut farm needed homes for the workers and a school for their children. A similar pattern emerged. The needs of the employees were a priority for George Shaw and education was top on the list. When Dale's wife, Betania, wanted to start a school, he supported it "100 percent." Even though he lived away from the Melbourne area for twenty years in Costa Rica, he always planned his trips up to Melbourne to coincide with being able to attend the Florida Tech board meetings. Recently while at a 50 year Florida Tech celebration, Shaw was asked by his daughter, Jerrie Hixon, if he had ever missed any of the Trustee meetings in the 50 years. He said, "Maybe one or two." The Radiators, as the former Radiation employees call themselves, speak of him as a colorful personality whom they respected highly as the foundation of the company's electrical engineering strength and inventive culture. The Florida Tech fellow Trustee members have known him for his strong opinions, yet true love for the school and how it can serve the community.

George's daughter, Jerrie Hixon, said that several years ago she accompanied George to the 50-year celebration of Radiation/Harris being in Palm Bay. It was the first time George had been back in those buildings since he resigned in 1966 after the merger with Harris. She said he enjoyed the breakfast program and slide show of some of the early days, and when a reporter asked him before the program started who he was, he boomed out, "I started this damned company!" That's our George!

George Shaw died June 15, 2010 at the age of 88 – but the bright side of that would be that he probably died the way he would have wanted to – peacefully, in bed with two flat panel TVs going and a brand new minicomputer having its battery charged so he could begin playing with it the next morning – George still loved gadgets and new technologies, even at age 88! But as those who worked with and for George know, he was far more than just a technology buff – he was a true visionary on many fronts. The community has largely forgotten the many positive impacts that George had on our sleepy little beach town – not just Radiation, Inc., but look at the merged cities of Melbourne and Eau Gallie, look at Holmes Regional Hospital, look at Florida Institute of Technology– and the list goes on.

Bill Dodgson
From the 1955 brochure:
William W. Dodgson, Jr., Vice President in charge of Contracting and customer services. He and his Division are responsible for contractual negotiations and direct the progress and completion of contractual commitments... acting as liaison between our engineering sections and the customer, as well as the government contracting officer or the contracting personnel of private industry.

Prior to joining RADIATION, INC. Mr. Dodgson headed his own corporation, having branch offices in Dayton, Ohio and Washington, D.C.

Mr. Dodgson's contractual experience was obtained at Wright-Patterson Air Force Base, where he held the position of Purchasing Agent in the Research and Development Procurement Division.

He is a graduate of Miami University, Oxford, Ohio, majoring in Business Administration. His military service included the Air Corps, Flight Cadet Training Program. Later he attended courses in Radio, Radar and advanced Electronics and participated in development assignments, in radio controlled aircraft at Wright-Patterson Air Force Base.

Bill was born December 19, 1924, in Dayton, Ohio. He attended Butler University before serving in the Army Air Corp in WWII, and after discharge he graduated from Miami University (Oxford, Ohio) with BS degrees in mechanical engineering and business administration.

A. B. Amis writes:
William Dodgson was VP in charge of customer services. Actually his role was marketing, but no one wanted to use the term in conjunction with Government contracting. His approach to marketing was

more personal than professional – "good timing" potential customers, promising them jobs, whatever it took to secure their business. This type of salesmanship was doubtless helpful in the early days when Radiation really didn't have much to sell, and helped the company corner the PCM telemetry market in the late 1950s, but as Radiation's reputation grew there was less and less need for this kind of salesmanship.

Bill was a very frustrating guy for an engineer to work for — if you disagreed with him about something, instead of discussing it his standard response was, "You're right, but you're wrong" and then he'd go ahead and do whatever it was his own way. I can recall a meeting once between Bill and George Shaw back in the mid-1950s to set internal priorities on contracts then in progress. We were late on basically all of our early PCM telemetry contracts, and when George proposed that we name the AC Sparkplug contract as number one priority Bill agreed. Then when George proposed that the Holloman Sled contract be named number two priority, Bill disagreed – insisting that this contract also be defined as number one priority. George tried explaining to Bill that only one job could be number one, and then the next job number two and so forth on down the line. Bill said he understood that, but that nevertheless the Holloman job also had to be number one – and so it went on down the line, with Bill insisting that each job they considered also had to be number one priority. Bill's credo was "stick with me and you'll be wearing diamonds," and I can remember once when that came back and almost bit him in the rear – at a meeting of his contracts/marketing staff he'd told all of us "there's no reason why everyone in this room can't be making $20,000 a year by next year." A year or so later, after he'd been forced to fire our Southwestern representative for spending his time in a Mexican jail instead of representing us, the rep sued the company, claiming that Bill had promised him a salary of $20,000 per year. The case probably never came to court, but controversy like this always seemed to swirl around Bill.

Bill was one of a group that broke away from Radiation in 1961 to form Systems Engineering Labs (SEL) in Fort Lauderdale. Some years later I asked Homer if he missed Bill. "Well, A.B., it was a lot more interesting when Bill was here, but I sure do sleep better now – not having to worry about possibly going to jail!"

In Fort Lauderdale he became a member of the Mormon Church and later, after moving to Utah, became a respected elder in the church. He moved to Utah in 1963 and worked for Thiokol Chemical Corporation before starting his own automobile dealership, which he ran for 25 years. Bill was a world traveler and hunter, and history buff. He visited and hunted in more than 70 countries, and traveled more than 20 times to Africa.

Bill died in Utah in 2007 of Parkinson's disease.

John Boone
The 1955 Brochure says:
John W. Boone, Treasurer, received his BS (BA) degree from the University of Florida in 1938, majoring in accounting and graduating with high honors... earning membership in Phi Delta Phi and Beta Alpha Psi.

Mr. Boone is a certified public accountant. He served as Assistant Secretary at McKesson & Robbins, Inc. prior to the war, and as Chief Accountant of Groover-Stewart Division. Currently he is President of Central Florida Chapter of Florida Institute of Certified Public Accountants. Mr. Boone served with the U.S. Navy from 1942 to October, 1945 in the Atlantic and Pacific Theaters and returned to inactive duty as Lieutenant Commander U.S.N.R in October, 1945.

John was born in Michigan in 1914, and grew up in Jacksonville, graduating from Robert E. Lee High School. He served in WWII as a commander in the US Navy. After his discharge, he practiced in Orlando where he represented many citrus operations. In 1956, John joined the founders of Radiation Inc. as Chief Financial Officer and Treasurer. He died in 1999 at the age of 85.

The "Missile Brochure" listed 57 people on the engineering staff, so we have to be selective of those we mention here. I have chosen some who I knew personally.

A B Amis
From 50's brochure:

A. B. Amis III, Engineer, received his B.E.E. degree (cum laude) from Georgia Institute of Technology. His experience covers design and development of airborne antenna systems, pattern measurements and antenna matching, ECM antenna systems, ECM system studies, development of UHF and microwave receivers, I.F. amplifiers and micro-wave components. Since associated with RADIATION, INC. he has been assigned to the development of high frequency measurement devices.

A. B. is an overall nice guy, technically smart, with wide interests. He has had a period running an RV rental business, sandwiched with engineering periods at Radiation and Harris. (As I write this, he is on an extended road trip in his own RV.) He is an expert in growing and carving gourds. He is a talented writer -- be sure and read his essay in the Personal Histories section. He contributes great historical knowledge and encouragement to the Radiation Old-Timers group.

James Coapman
From the brochure:

James W. Coapman, Project Engineer, received his B. S. degree in Physics from Union College, and served as a naval electronic officer. As a graduate assistant at the University of Rochester for two years, he was engaged on the synchro cyclotron project. He was later connected with Bendix Radio where he served as Project Engineer in charge of a radar systems development, also in charge of a missile guidance study and the design and development of missile guidance equipment. Mr. Coapman's experience at RADIATION, INC. includes design and development of Servo equipments, a wide range Spectrum Analyzer, a Servo-controlled vehicle, airborne and ground telemetering equipment, and aircraft reflection measurements, instrumentation and flight testing. He is a Senior Member of the Institute of Radio Engineers.

Jim is one of the very early employees, but also left Radiation early to start his own company, Missileonics, Inc., where I worked for him from 1957 to 1959. Missileonics did not last long and never made lots of money, but Jim made the most of the perks of a family-owned business -- he had a company car and a company boat. He and his wife ate out almost every evening -- often at the Bahama Beach Club, and regularly invited me and my wife. I learned a lot about after-dinner drinks from the Coapman dinners. He was a joy to work for.

Mel Cox
From the brochure:

Marcus L. Cox, Engineer, received his B.E.E. degree from Georgia Institute of Technology in 1951. His experience covers design and development of aircraft UHF transmitting and receiving antennas. His antenna and propagation instrumentation developments included special model aircraft test devices, extremely broadband antenna, high accuracy R. F. wattmeters, and high frequency pattern recorders, radar reflection recorders and other high frequency instrumentation. He is a licensed commercial pilot and holds amateur and professional radio licenses. Mel had received training as a Naval Officer.

Mel had many diverse interests and loved to ask questions and discuss things. A. B. Amis was a close friend of Mel's dating back to college days. As a matter of fact, Mel's interview at Radiation got them both hired. A. B.'s remembrance of Mel in the Personal Memories section is a must-read. As an example:

"Let's see what other silly stuff Mel did. Well, one of his prouder accomplishments was being able to touch the tip of his nose, or his chin, with his tongue – and having the dexterity to tie a knot in a cherry stem with his tongue." Sailing was an important part of Mel's life after retirement, and he even got into "the arts" briefly after getting hold of equipment for forming and firing ceramics.

Mel died in early 2002 following several years of declining health – precipitated probably by the heavy smoking that he was seemingly never able to give up for long.

Bill Eddins
From the brochure:
William T. Eddins, Jr., Engineer, received his B.E.E. degree from Alabama Polytecnic Institute. Under a graduate assistantship, he attended Florida Stare University where he received his M.S. degree. He served as a design engineer on electrical equipment at Douglas Aircraft Company, Santa Monica, California, working on modifications of C-54 and DC-6 aircraft, on design of DC-7 aircraft and on a study program for projected DC-8 jet-propelled passenger aircraft. Since joining RADIATION, INC., he has worked on the AKT-6/UKR-1 telemeter project. His experience covers mechanical layout of packaged circuitry, plus production engineering of complex electronic assemblies. He is a member of the Institute of Radio Engineers.

Bill was an expert circuit designer, responsible for many of the analog circuits in early airborne systems. He cultivated a thick southern accent and had a million funny stories and observations and was not shy about expressing his opinion of foolish behavior by others.

Dave Howard
From the brochure:
David C. Howard, Senior Engineer, received his B.E.E. and M.S.E. degrees from the University of Florida. While at Florida he served as student assistant in the electronics laboratory and later as graduate Research Assistant with the vacuum tubes project. His experience covers development of analog computers, D.C. amplifiers, phase shift oscillators, and wide band IF amplifiers. He has designed and developed an 18 channel airborne PAM/FM telemetering system, also he has had responsibility for redesign and modification of the AKT-6 telemeter. This covered design and development of blocking oscillator, multivibrators, stable audio carrier amplifiers, video amplifiers and pulse circuitry. Subrniniaturization of packaged and associated circuitry, also environmental testing and flight test of the AKT-6/UKR-1 system. He is a member of the Institute of Radio Engineers.

Dave was a soft-spoken but skillful logic designer and project engineer. He was an early devotee of foreign sports cars and neighbors could tell when he was approaching the neighborhood because his ignition wires produced interference on their TVs. Dave died young, of a heart attack while playing tennis.

Ralph Johnson
From the brochure:
Ralph A. Johnson., Project Engineer, received his B.E.E. degree from the University of Florida. He served for three years in the Engineering and Industrial Experiment Station sferics research project. Development of electronic equipment designed for sferics study resulted from this basic research and data collection. He remained at the University of Florida pursuing advanced studies and serving as Graduate Assistant until he joined RADIATION, INC. His experience at RADIATION, INC. covers design and layout or certain portions of the Interim Pulse Timing System for AFMTC. Responsibility for Pulse Circuitry Development of the A.N/UPM19 (XA-2). Prime responsibility for development of subminiature complex pulse generator utilizing transistors, plus the TS 624 (XA - B)/APW Pulse generator. Also our present line of Servo-meters and Super Regulated Voltage and Current Standards. He is a member of Phi Kappa Phi and the IRE.

Ralph was employee #19, hired early in 1951. Ralph moved into management positions and was sent to Harvard Business School where he received a Masters in Business Administration degree, and was later named Director of Products Division.

Bob Marquand
From the brochure:

R. Marquand, Senior Engineer, received his B.S.E.E. and M.S.E.E. degrees from the Georgia Institute of Technology. After receiving his B.S.E.E. degree he joined the Sandia Corporation as a staff member engaged in design and development of vertical velocity calibration units, precision timers and range circuitry. He later joined the staff of the Georgia Tech Research Station, where he was engaged in research on underwater sound apparatus. His experience also covers development of analog to digital convertor systems and pulse code modulation. He is a member of the I.R.E.

Bob was a section head when I was first employed. I recall him as being very helpful to a confused new hire.

Parker Painter
From the brochure:

Parker Painter, Instrumentation Division Manager, received his B.S.E.E. degree from Massachusetts Institute of Technology. He has been engaged in Radio and Electronics with Commercial and Military organizations for thc past 12 years. At Melpar, Inc. he was Senior Engineer on the AN/AKT-6 and AN/UKR-1 telemetering system; in charge of portable ground direction- finding receivers and multichannel instrumentation amplifiers; and assigned to problems in compressed bandwidth speech transmission. After joining RADIATION, INC. in March, 1951 as a Project Engineer he was in charge of the following: AFMTC Interim Timing System, Matador missile telemetering group; AN/UPM19 (XA-2) radar test set; design of a pulse microwave S-Band signal generator; redesign and development of the AKT-6/UKR-1 telemetering system; UPM-17 wide range spectrum analyzer; UHF wattmeter; URM-57 video dummy load; special low-noise L-Band radar receiver; and study contract on electronic countermeasures equipment. At present, Mr. Painter is in charge of the Instrumentation Division of RADIATION, INC. He is a member of the IRE.

Parker worked in Orlando most of his Radiation career. He formed one of the earliest spin-offs, Dynatronics, Inc., in 1957.

Hans Scharla-Nielsen
From the brochure:

Hans Scharla-Nielsen, Senior Engineer, received his B.S.E.E. degree from the University of Florida. Previous to joining our staff he was employed by the General Electric Research Laboratory, engaged in vacuum tube development; he later worked on electronic counter measures and mono-pulse radar development until joining the staff of RADIATION, INC. His recent experience covers development of VHF and SHF components including wave meters, gauges, oscillator and R.F. Cavities, variable attenuators, antennas, VHF and SHF receivers, Electronic Countermeasure Systems Study. He is a member of IRE, AIEE and Sigma Tau.

Hans was born in Westchester County, N.Y. He came to work at Radiation in about 1953. He was a member of Holy Trinity Episcopal Church, and was past commodore of Melbourne Yacht Club. Hans was a versatile and talented engineer, and served as Chairman of Radiation's Patent Committee for many years. He held patents in fields including adaptive power control of satellite communications and a radio control systems for trains.

A.B. Amis writes:

While Hans' office was a disaster to look at, with papers piled everywhere, he was actually an engineer's engineer -- meticulous about filing important documents so they could be retrieved easily, and scrupulous about making daily entries in his engineering notebook. Every day's mail brought catalogs, and Hans kept a list of the many different permutations of his names that they were addressed to. He was an inveterate pipe smoker (he had a <u>huge</u> pipe that he used when borrowing pipe tobacco) and was also "fond of the grape". He actually had three different careers at Radiation/Harris – "Mr. Radio Telemetry" in the 1950s-1960s, "Mr. Satellite Communications" in the 1960s-1970s, and "Mr. Railroad Technology" in the 1980s. Two of these careers were interrupted by this fondness for the grape and the third by throat cancer that finally killed him in at age 65 in 1992.

John Searcy

From the brochure:

J. H. Searcy, Jr., Engineer, received his B.S.E.E. and M.S.E., (Thesis-Electronic Methods for the measurement of small D.C. Currents) from the University of Florida. His past experience includes Navy radio repair and maintenance, also worked as laboratory assistant at RCA. His work at RADIATION, INC. has included dielectric amplifiers, magnetic tape recording, and super regulated power supply developments.

John was a talented circuit designer. He was largely responsible for the design of the Radiplex Low Level Multiplexer. He was a bachelor of simple tastes and would sometimes let his paychecks accumulate in his desk drawer until Accounting would finally call and remind him to cash them.

Earl Smith

From the brochure:

Earl F. Smith, Assistant Project Engineer, received his B.S.A.E. and M.S.E.E. degrees from the Massachusetts Institute of Technology. His past experience covers automatic control of aircraft tracking and control at MIT Instrumentation Laboratory, and theoretical investigation of various servo systems for automatic control of aircraft at NACA, Langley Field, Virginia. Recent experience covers development of precision missile tracking equipment, design and development of servo- systems and computers for automatic control applications, flight test instrumentation and radar systems. He has written numerous technical papers in the field of instrumentation and is a member of Gamma Alpha Rho.

Earl was born in 1920 in Granite Falls, NC, and joined the CCC in 1937 before finishing high school to help support his large family. Afterward he worked for the US Forest Service for two years before enlisting in the US Army Air Corps in 1941, becoming a flight instructor in B-24 bombers in 1943. Leaving the service in 1945 Earl was able to talk his high school superintendent into awarding him a diploma based on his military service so he could attend MIT, where he earned BSAE and MSEE degrees. After working as an engineer at Radiation for several years he went back to Cal Tech and the University of Michigan where he earned a PhD in statistical communications. At Radiation he was a skilled engineer and scientist, who always had time to answer your most arcane questions.

Earl died of lymphoma in 2005 in Young Harris, GA.

The earliest known group picture, taken in front of Building 1 at the Melbourne Airport in 1953. The best available identifications are labeled on the picture or in the numberec list below. Appogies for any errors.

Who's who— in 1953

#	Name	#	Name	#	Name
1	Jerry Newell	35	John Spooner	68	Bob Lebo
1(alt)	Jim Jenkins	36	George Wicker	70	Seth Stevens
2	Bill Pierpont	37	Dick Morse	71	Don Graham
3	Nick Raber	38	George Hedman	73	Joe Petras
4	Ralph Johnson	38A	Dan Parker	75	Dave Howard
7	Bob Goldman	39	Pierre Toulon		
8	Frank Sella	41	H.F. VanLeuven		
9	Parker Painter	43	Joe Detweiler		
10	Charlie Keith	44	Bob Dryden		
13	Noel White	46	Hans Scharla-Nielsen		
14	Ed Dorsett	49	Bill Corry		
15	Roy Gilbreath	51	Jimmy Maxwell		
16	Bu Chen	52	Paul Beckwith		
18	Joan Porter	53	Bob Bishop		
20	Jim Coapman	54	Jim McKee		
23	Vince Soltero	55	Pete Robeson		
24	George Wicker	56	Harry Siler		
26	Dean Ashwill	57	Ann McDonnell		
27	Jimmy Murray	59	Teddy VanEghen		
28	George Anderson	61	Hal Gettings		
29	Marilyn ?????	62	Jessie Ferguson		
33	Jack Petersen	66	Jim Danner		

From the October, 1965 Radiation Ink. These would be people employed by 1955.

58 EMPLOYEES HAVE 10 YEARS OR MORE OF SERVICE

1, Julian Scott. 2, Ralph Johnson. 3, Roy Ellis. 4, Joe Detweiler. 5, Bob Lebo. 6, George Shaw. 7, Don Flick. 8, Oscar Seaberg. 9, Bill Hogg. 10, Ken House. 11, Dick Addy. 12, Tom Wicker. 13, Bob Marquand. 14, Leo Hosley. 15, Al Carnes. 16, Harold Caudill. 17, Carl Satterlee. 18, Bill Pierpont. 19, Jim Mathis. 20, Jesse Ferguson. 21, Bill Eddins. 22, John Mazur. 23, Ollie Wiseheart. 24, Jim McKee. 25, John Spooner. 26, Earl Smith. 27, Robey Green. 28, Bonzie Barfield. 29, Al Johnson. 30, Larry Bradbury. 31, Oscar Berning. 32, Walt Florin. 33, Wally Silver. 34, William Kercher. 35, Kay McCall. 36, Julia Ruhlander. 37, Marge Coker. 38, Gladys McClung. 39, Edna Tutwiler. 40, Arthur Maurer. 41, John W. Boone. 42, Travis Willis. 43, Ed Weiss. 44, Bill Corry. 45, Howard Yates. 46, Norma Kasten. 47, John Cahill. Absent when this picture was taken are Homer R. Denius, Hans Scharla-Nielsen, A. B. Amis, Jack Fifield, Phil Derrough, Harry Fisher, Ray Penn, Harry LeGrand, Reed Barnett, Wayne Williamson and John Diaz.

Radiation Ink

Page 3

From May 1968 Radiation Ink

The Cape Canaveral Chapter of the National Association of Accountants (NAA) was selected in competition as the best chapter of its size in the United States during 1967. Among the chapter members who made this honor possible were several accountants from Radiation.

By placing first among 42 chapters nationwide, the local chapter won the coveted Warner Trophy. Chapters were judged on performance in 10 areas including membership, educational activities, attendance, and manuscript.

Among the Radiation accountants who are members of the Cape Canaveral Chapter are, from left to right, Bill Nibler, Tom Moldenhauer, Ernest Ploegger, Roger Smith, Bill Costill and Millard Johnson.

Costil and Johnson were chapter officers during the award winning year.

The Radiation Patent Committee met April 19 to present awards to employees who have had an application accepted for filing or have been issued a Letters Patent recently. Recipients are, from left to right, B. Huston Singletary, Bernard L. Lewis, Misha I. Kantor, Billy H. Thrasher, Frank A. Perkins, James A. Proctor, Eugene B. Stewart and Hans Scharla-Nielsen.

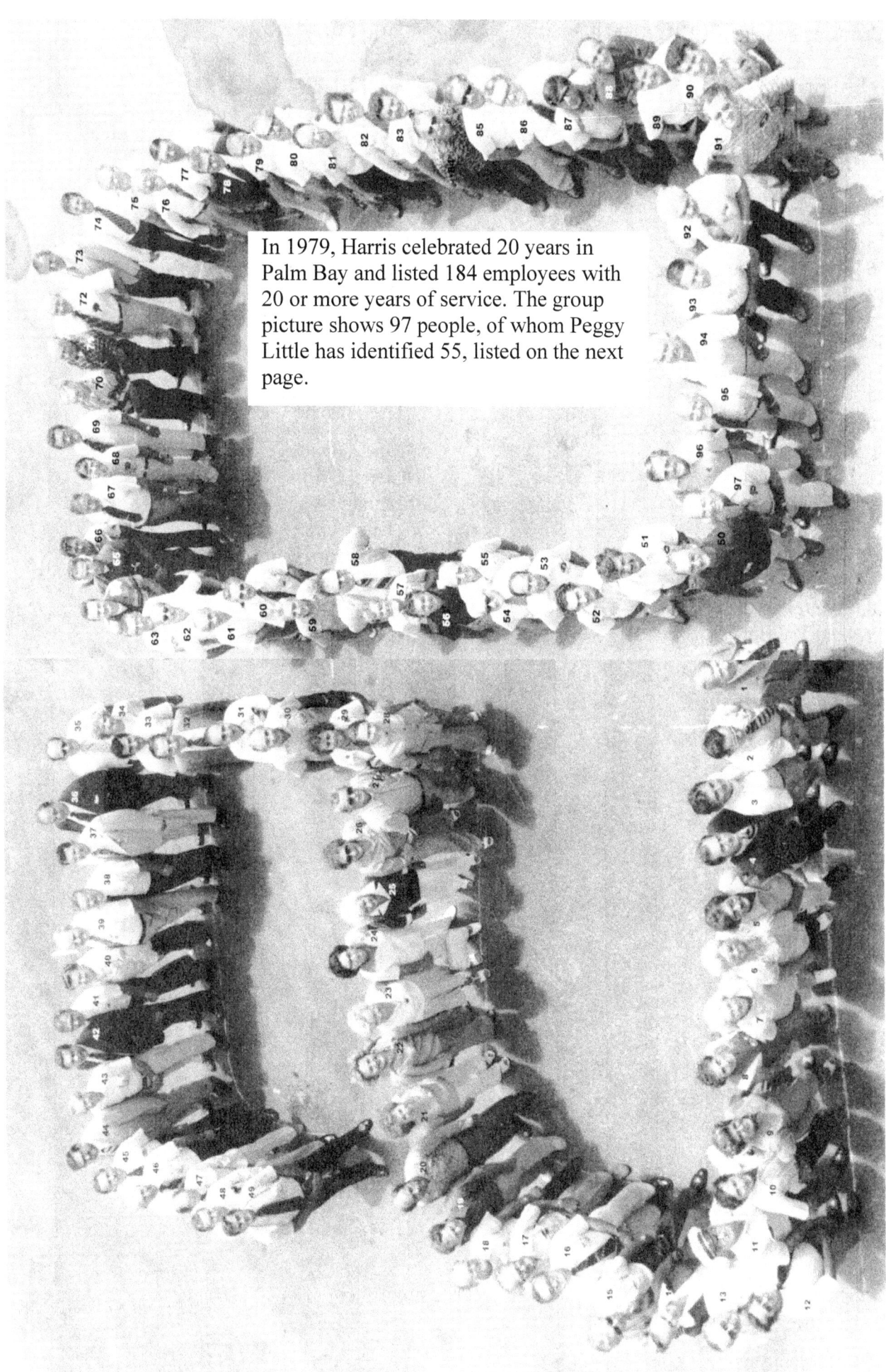

In 1979, Harris celebrated 20 years in Palm Bay and listed 184 employees with 20 or more years of service. The group picture shows 97 people, of whom Peggy Little has identified 55, listed on the next page.

Identified people in 1979 Group Picture

1. Bill Haney
3. Bob Underill
4. Wayne Sumner
5. Bill Taylor
6. Wynelle Harris
10. Gantt Hamner
15. Bill Kercher .
16. Jack Tomlinson
18. Jim Perkins
19. Gene Laflotte
22. Anne Bouvier
25. Alma Rinehart
27. Joel Warren
28. Kathy Quick
30. Sol Smith
31. Larry Bradbury
33. Tom Little
34. Collette Waltman
37. Bill Quinlivan
44. Al Gerstle
48. Lonzo Parrish
49. Fred VanBever
50. Tom James, Jr
51. Richard Ott
52. Mike Rabits
53. Bill Stankos
54. Leo Hosley
55. Bill Corry
56. John Skolnik
58. Bill Hicks
61. Andy Sitton
62. Neil Brooks
64. Phil Christensen
68. Frank Perkins
69. Mel Cox
70. Ed Glover
71. Dan McRae
72. Bud Mills
73. Hans Scharla-Nielsen
74. Bill Tolley
75. Tom Sheffield
77. Curt Gallagher
78. John Spooner
83. Jim Townley
84. Jerry Grillo
85. Julian Scott
86. Don Shingler
89. Bill Morrissette
90. Bill Ludwick
91. Scott Broadway
93. Stan Stocker
94. Ralph Haack
95. Ray Penn
96. Don Flick
97. Earl Smith

The Apollo/Saturn Data Processing System built for North American Aviation, and delivered to Downey, CA in the summer of 1964. Identified people, starting with top row, left to right:

Howard Thrailkill, Gene___, Unknown, Unknown, Al Heitman, Unknown, Dick Harvey, Bobby Kemp, Frank Lewis

Herb Newkirk, Unknown, Larry Beyer, Ed ___, Unknown, Jack Snow, Bill Arbuckle, Unknown, Russ Latch.

Unknown, Bob Haddock, John Lemasters, Ed Hersberger, Walt Fredrickson, Sinclair Fredrick, Unknown,, Ryan Heath, Ed Glover.

Mark V (TSC-54) delivery photo. November, 1967.

Jack Hartley presents an award to Jim Danner

Some olld-timers at Frank Perkins' 25th year party in 1984. from left to right: ?, Bob Marquand, Larry Gardenhire, A. B. Amis, Earrl Smith, Bernard France, obscured, Guy Pelchat, Bill Buchanan, Harland Carthures, Frank Perkins, Ray Penn(?), Larry Klingler (?), Sandra Cope, Mel Cox, Jane Ford, two obscured people, Sam Boor, Dan McRae, Tom Losson, Ed Glover.

Radiation's 15-year-old Aero Commander recently received a new paint job and interior decorations. The twin engine, five passenger aircraft has flown 1,654,-000 miles. Shown with the aircraft are the company's staff of pilots, from left to right, Terry Daubenspeck, Robey Green, Willard Van Dusen, Joe Latorre and Jim Clevenger. Jim Walsh was absent when this was taken.

Radiation Pilots from about 1971.

Chapter 3

Personal Memories

Here we present, in alphabetical order, some personal memories of early days at Radiation of and by a variety of Radiators, including experiences of life in the community.

A.B. Amis's Memories

George Shaw says Radiation, Inc. co-founder Homer Denius was prone to "shoot from the hip" in hiring folks in Radiation's early days – maybe someone he'd met on an airplane would show up a short time later as Chief Engineer. And I can certainly believe that – I was hired without even an interview. Well, actually, Mel Cox interviewed for both of us, but let me go back to the beginning.

Even though I was born and raised in one of the poorest counties in a state (MS) perennially ranked at the bottom of any list in terms of literacy, income, or whatever, my timing was charmed. World War II ended as I began my senior year in high school, but with the draft still hanging over my head I enlisted in the Navy on a two year hitch, with a guarantee of going to electronics schools for most of that time – which set me off on an electronics career path and paid my way through college via the GI Bill. Finishing Georgia Tech in December of 1951, I, along with two close friends and lab partners (Mel Cox was one of them), accepted our highest paying job offers, which were with a small electronics R&D company in Evansville, IN. One hot summer and one cold winter in Indiana were all it took to convince Mel Cox and me that we wanted to work in sunnier climes, so I responded to a squib of an ad in the back of the IRE professional magazine for another small R&D company in Florida (Radiation, Inc.) and made arrangements for Mel to visit Melbourne and look them over while on a Christmas 1952 trip back to his home town of Orlando. Radiation ended up hiring both of us on the strength of that one visit/interview by Mel, with a reporting date in early February 1953 and a starting salary of $4000 per year.

At the time we reported for work, the entire Radiation work force of about 70 people was housed in a single building–a former Navy barracks building immediately north of the present day Trailer Haven Recreation hall. Maintenance manager Jimmy Murray and electrician Walt Florin were renovating a second building just south of the present airport terminal building, and this became known as Building 2, with many of the administrative and engineering functions moving into it a few weeks later.

Mel and I were assigned to work on Project 1022, which was a CPFF contract with the Air Force to develop a broadband RF wattmeter. The proposed sensing element was to be a thin ceramic tube coated with a film of resistive alloy material on the outside, and with a thermocouple inside to sense temperature changes as the RF energy heated the resistive film. We were assigned an office/lab, a vacuum pump fitted with a large glass Bell jar, and a technician, Joe Detweiler, who didn? know any more about vacuum deposition than we did. We got busy trying to vapor deposit a 50-ohm film on a ceramic tube, and I can remember that we all three vacated the room the first time we started pulling a vacuum with that pump ·fearful that the Bell jar would implode as the diffusion pump began scavenging the last few molecules of air and approaching a pure vacuum in the jar. I believe it was Phil Derrough, who had worked in a college chemistry lab in Ohio before joining Radiation, who later reassured us by pointing out that even with a pure vacuum, the pressure on the glass would really only be about 15 psi. We contrived all manner of Rube Goldberg schemes for vaporizing resistive alloy metals, while at the same time rotating the ceramic tube via magnetic coupling through the glass jar and trying to measure the resistance of the film via slip rings contacting silver bands painted onto the ends of the tube. Probably the only thing of any value that came out of several months of futile efforts to create the sensing element was that Mel and I learned how to make entries in our engineering notebooks and write weekly progress reports to our boss, who might have been Parker Painter.

While Radiation had been founded primarily with the expectation of doing radio telemetry projects, several other early projects deviated from that plan, like our wattmeter project. I can recall that a guy named Jim Jenkins was working on a "dielectric amplifier" project, Charlie Keith on a multi-stylus flight data recorder, Ed Dorsett on an L-band radar receiver, and Ralph Johnson on something called the UPM-19 Test Set that went on to production later when Fred Cullman established a small manufacturing operation in Orlando. After our wattmeter project had fizzled out, I believe Mel was assigned to work for awhile with Hans Scharla Nielsen on a high power amplifier to boost the 2 watt output of our telemetry transmitter, and then later on radar reflectivity projects that became his area of specialization for quite a few years. Incidentally, when you read about stealth aircraft now, think about Mel and his crew of Wayne Williamson, Ronnie Evans, Pete Young, and others out at "80-Acres", who were working, even then in the 1950s on Air Force projects aimed at reducing radar cross sections of aircraft -- think "stealth".

When our wattmeter project closed, probably in mid-1953, I was assigned to work on Project 1023, which was an S-band PTM/AM telemetry system. Someone else, probably Melpar, had done earlier development work on the system for the Air Force, but Radiation did extensive re-design and actually manufactured equipment that deployed to the field for the Operation Teapot atomic bomb tests in Nevada in the mid-1950s. My initial job was to design a low noise 60 mc amplifier to interface with a 2200 mc balanced mixer, and after completing that I was given the job of picking up the pieces on a 30 mc IF strip development that had been begun by Harry Siler. I believe that I reported to Bob Dryden, who was in charge of the 2200 mc receiver and transmitter developments, while other folks like Bob Bishop and Dave Howard reported to George Anderson, who was in charge of the PTM signal generation and decommutation developments. Some of the technicians working on Project 1023 were Don Graham, who was my technician in IF strip design, Jim Danner, who I believe was working with Dave Howard and Bob Bishop, and Ken Pohlman, who was working with an older reformed alcoholic engineer named Al Campbell on the 2200 mc transmitter development. A mineral oil laxative called Nujol had been discovered to have exactly the right properties for conducting heat away from the 2200 mc airborne transmitter, and Ken evidently raised quite a few eyebrows when he was repeatedly dispatched to a local drugstore with orders for several gallons of the stuff. And the story is told on Al Campbell that once on a trip somewhere, Homer Denius engaged him in a conversation about his earlier drinking. "Al, I understand that you used to drink quite a bit." "Yes, Homer, I did." "Well Al, how much did you drink?" "Oh I drank quite a lot Homer." "Well Al, did you drink as much as a pint of whiskey a day?" "Hell, Homer, I spilled more than that!"

Upon arriving in Melbourne, we immediately bought a brand new 2-bedroom home in Magnolia Manor for $8500, with monthly mortgage payments of something like $48 per month. No air conditioning of course back in those days, and in fact the home didn't even have any provision for heat until the next fall when we finally got around to buying a kerosene powered space heater fed from a 55 gallon drum mounted atop several concrete blocks out back – that was the standard formula for most of the homes in that area. Living within a block of us there in Magnolia Manor were about a dozen Radiation employee families, and employees who had arrived in Melbourne a year or so earlier were likewise concentrated in the Loveridge Heights section of Eau Gallie, which was a separate town in those days. The unmarried folks were scattered all across older sections of town in apartments, but bachelor Gantt Hamner was enterprising enough to lease a home in Magnolia Manor and take in several roommate.

Gantt was also sort of a "sparkplug" for promoting fishing trips and water-skiing fun in the Indian River (which was clear enough back in those days that you could see the bottom in several feet of water). He once talked me into wading all the way from Melbourne to Indialantic in knee-deep water beside the causeway at night, carrying a lantern and looking for flounder. No flounder, but we stepped over lots of hermit crabs and stingrays. Gantt had a boat with a ten or fifteen horsepower Johnson kicker, and lots of us learned to water-ski behind that boat on Sunday afternoons – usually on the north side of the causeway near Indialantic, which was a favorite gathering place for family picnics, etc. Parker Painter showed up on the causeway one Sunday with a boat powered by a new 25 hp Johnson, which was top of the line at that time, and not to be outdone, Ralph Johnson showed up shortly thereafter with a 25 hp Evinrude – only he had neglected to attach it to the boat with a safety chain and ended up having to do a minor overhaul after it fell in the river.

The entire population of Melbourne at the time we moved here was probably less than 3000, and virtually all of the businesses were concentrated along New Haven Avenue, east of Ruth Heneger School -- which was the only school in Melbourne at that time – grade school and high school combined. The businesses consisted mainly of Turner's and Coleman's department stores, Dennis Medvine men's' clothier, Lovett's grocery store, the Van Croix movie theater, Harry Goode's Outdoor Shop, a handful of drugstores, barbershops, and bars, plus several law offices and doctor's offices. I well remember when the U-Totem convenience store opened on US-1 and Ballard Drive in Eau Gallie – before that you were out of luck if you needed a jug of milk or loaf or bread after Lovett's closed at 5:00 pm. If you really wanted to do much shopping, you were looking at a trip to Orlando – which wasn't always so easy, especially if Hwy 192 was underwater out around the St. Johns bridge – as if frequently was during the rainy season. There were very few homes over on "the beach" at that time, and Hwy. A1A southbound ended at the old Coast Guard station not too far south of Melbourne Beach – just sandy ruts made by a few determined fishermen owning beach buggies from there on down to Sebastian Inlet.

Back to my working career at Radiation: after my Project 1023 IF strip designs were complete and documented in late 1953, I was briefly assigned to Project 1037, a "Countermeasures Intercept Receiver" study that really went nowhere, and then I was transferred back to Project 1023, which by this time was converting our earlier "breadboard" designs into manufactured equipment for shakedown testing at Eglin Field in advance of being deployed to the field to support Operation Teapot. This would have been late 1954 to early 1955.

In late 1954 I was temporarily dispatched with one technician, a camera, and some numbered adhesive strips to Wright Field in Dayton, Ohio, to prepare a prototype PCM telemetry system for shipment to Radiation. I believe Melpar had probably done the initial development work on the PCM system for the Telemetry Lab at Wright Field, and Radiation had evidently been awarded a contract to do follow-on development. There was no documentation, so our task was to make photographs and attach the numbered adhesive strips to every wire and terminal as we disconnected them so it could be re-connected after arriving at Radiation. This was the beginning of Radiation? work in PCM telemetry a field that it would later dominate

In early 1955, I was encouraged to transfer into Contracts Department (with a big raise in pay to $10,000 per year) to work as Assistant to the VP Contracts, Bill Dodgson. I can see now, in retrospect, that my transfer into Contracts in 1955 was probably intended as cross training, hopefully preparing me to move up in management – a career path for which I really had no ambition. Jack Hartley was afforded a similar opportunity and ran with it. While Contracts Department did, in fact, monitor contracts in process, Bill Dodgson's main thrust, and mine also as his assistant, was in marketing – although you were forbidden to use that word in connection with government contracting. Radiation was, at that time, operating at about a $2M per year annual sales rate, but Homer was clamoring for a $10M contract backlog and beginning to toy with bringing in business (Lou Clark) and technical people (a small research group) from outside the company. My first duties in Contracts/Marketing were to try to develop a systematic way to handle Requests for Quotation and subsequent proposals. Up to that point an RFQ might come into the house and be handed, informally, by George Shaw to some Project Engineer to look at – and then get "lost" until too late to make a bid. Formal procedures were developed for routing and tracking RFQs and for providing subsequent Contracts/Marketing participation in the review and editing and publication of proposals. I didn't realize it at the time, but thus began a whole career for me of late nights and long weekends working against tight proposal deadlines.

Bill Dodgson's approach to marketing was more personal than professional – "good timing" potential customers, promising them jobs, whatever it took to secure their business. Using that marketing approach and using the 1047 PCM telemetry contract as a cornerstone, Radiation was able to parlay that initial PCM contract into orders from the Avionics Lab at Wright Field for a PCM flight data recording system, from Holloman AFB for a sled-borne PCM system, from Kirtland AFB for another PCM system for Special Weapons testing, another PCM order from A.C. Sparkplug who was working on an IRBM missile contract, and mercifully another large ($10M) CPFF order for PCM equipment from AVCO, who I believe was under contract to develop an ablative nose cone for Atlas ICBM missiles. Most of these jobs had been knowingly under-bid, counting on synergy between the several jobs to assist with development costs, but divergent technical requirements (some 8-bit and some 10-bit systems) and divergent delivery requirements (one-dash-one "brickbat priorities" on the missile-related programs) got in the way of lots of the hoped-for synergy, with the result that by 1957 Radiation found itself facing about $10M in cost overruns prior to award of the CPFF AVCO contract, which proved to be a "golden goose" in allowing us the financial means to go forward with needed PCM development. It hadn't helped matters, either, that Parker Painter had taken quite a few key technical people with him at this critical juncture when he concluded to leave his job as head of the Instrumentation Division in Orlando to form Dynatronics – the first of many spin-off companies incubated by Radiation.

The job in Contracts turned out to require more travel than I had any appetite for, and particularly so after the birth of our first child in 1956, and on top of that I had come to realize that I had basic "philosophical" issues with Bill Dodgson's style of marketing, so in early 1958, after nearly three years in Contracts, there was mutual agreement that I should transfer back into Engineering. Incidentally, Bill Dodgson himself was involved in a spin-off on a year or so later – joining Gus Randolph, Tom Sullivan, and several other Radiation engineering people in forming SEL in Ft. Lauderdale. I asked Homer, some years later, if he missed Bill. "Well, it was probably more exciting when Bill was around, but I surely do sleep better now, not having to worry about possibly going to jail!"

After leaving Contracts Department, I reverted to my then "specialty" of low noise IF strip development when I was assigned to Project 6151, which was a fixed price contract from Sperry Gyroscope to develop a mono-pulse radar receiver front end employing stripline microwave circuitry. I don't now recall much about the development other than it had very stringent requirements on gain of the IF strips, requiring careful selection and testing of large quantities of vacuum tubes and the use of feedback to maintain the outputs constant. I don't recall whom I reported to on this project, but probably ultimately to John downs, who had been hired from Sperry Gyroscope and made head of RF Division, probably largely on the strength of his having brought with him a Doppler Navigation Radar project with Sperry as the customer (Project 1130). Progress seemed to be lagging on that project, which had been staffed six months or a year earlier primarily with new-graduate engineers (Jay Fleming was one of them), so Project Engineer Warren Wiener had gone to his management with a request for some more senior people. Ed Dorsett, Walt Minnerly, and I were assigned to the project in late 1958, and perhaps Guy Pelchat also – although Guy may have already been working on the project – he'd worked on a similar project at Canadian Marconi before coming to Radiation. I'm fuzzy on this, but we may have started off designing the Doppler navigator around micro-miniature vacuum tubes, but transistors were just coming on the scene at that time, so the whole crew attended lectures given by Bill Eddins on transistor circuit design. I took on the task of trying to design a transistorized 30 mc IF strip, and I guess I must have come up with something sort of workable because the project progressed on to flight testing some time later -- but the flight tests ended abruptly when the landing gear on the test aircraft collapsed while the plane was taxiing, smashing much of our equipment which protruded from the bottom of the aircraft in a radome. Project 1130 was initially housed in Bldg. 9 at the airport when I joined it, but moved shortly thereafter to the Malabar test facility. I actually remember far more about the people that I worked with than about the technical details of that project. Ed Dorsett and I had already been singing together in a barbershop quartet, and after assignment to work together on 1130 we managed to corral a couple of other co-workers (Dick Ferguson and Bob Fletcher) to "harmonize" with us during lunch breaks or other times when the four of us found ourselves together with a little time on our hands.

Shortly after the untimely end of the Doppler Navigator project, my recollection is that new hire Harold O'Kelley and I shared a small office in the rear of old Bldg. 2, back near the stockroom, for six or eight months, where we went through a period of bidding on a number of small tracking antenna jobs. It's probably lucky that we didn't win any of those jobs, because I believe our bid price on every one of them was only in the range of about $40,000. Both of us probably reported to John Downs in something of a "staff" capacity at that time. I believe O'Kelley moved up to replace John Downs as head of the RF Division shortly thereafter, and I started to work for Harold in a technical staff capacity, along with Bill Beatty and Guy Pelchat – primarily writing proposals for large diameter tracking antenna systems, trying to leverage Radiation's earlier successes with the TLM-18 60-footer. That must have been about 1960, and was the beginning of what were probably the happiest years of my working career.

In 1961, Harold O?elley and I collaborated with RCA on a successful proposal to Boeing for a K-band communications link for the Air Force? Dyna Soar manned boost-glide orbital re-entry research vehicle ·whose design could probably be considered something of a prototype for NASA? space shuttle some decades later. The Dyna Soar program was, unfortunately, cancelled in 1963, but not before we had completed development of a high precision tracking antenna. Following Dyna Soar we submitted a successful proposal to NSA for a wide band, highly sensitive, precision tracking, 85 foot antenna system employing an agile X-Y mount, and while in negotiations for this contract the customer added on a requirement for a lower frequency 150 footer. (This would have been in the 1962 timeframe, because I recall that we got news of JFK's assassination after returning from lunch with NSA personnel here in Melbourne.) These turned out to be the Stonehouse/Bayhouse antennas, later installed in Eritrea to eavesdrop on soviet missile and spacecraft launches, but of course that was all highly classified at the time -- necessitating polygraph tests and a separate secure workplace for project personnel. As developed to be the usual thing on subsequent programs, my personal contribution mostly just encompassed the very intensive proposal/sales effort, followed by a 90 day "design plan" phase. "Y'all skin this one while I go out and help catch another one!" Actually, I could usually get away with just "floating" for a month or so to "recharge my batteries" until the next big proposal effort came along.

Along in this period we had submitted an unsuccessful proposal for a tracking antenna system for the Tiros weather satellite, familiarizing us with its frequency of operation orbital parameters, etc., so when we received an RFQ shortly thereafter from the Air Force for a quick reaction program to install two antennas in the extreme NE and NW corners of the US for tracking something with parameters seemingly identical to Tiros, Mel Cox and I jumped right on it. A key element of our proposed quick reaction solution was to mount the antennas inside inflatable radomes to spare them the expected harsh temperature and wind environments and thus allow use of "already designed" antennas. After winning the job, all project personnel had to receive "special indoctrination" Secret clearances, because the antennas were, indeed going to track the Tiros weather satellites, for military purposes, in clear violation of existing international agreements. I remember one meeting that Mel and I attended with the customer out in Los Angeles where an excitable young Major Charlie Croft was drawing something of a Pert chart on the blackboard, denoting all the things that had to fall in place almost perfectly in order for the schedule to be met, and despairing of it ever happening, when he was interrupted by slow talking old Colonel Dan Gray from Georgia. "Charlie, he said, I know what we've got to do -- we've got to HURRY!" That was the beginning of a succession of meteorological satellite contracts that possibly continue even to this day.

Mel Cox and Dick Baker had made some good friends in the Army's Satcom agency by good performance on the Score and Courier communication antenna projects, as well as another project whose name escapes me right now, for establishing a satellite communications "hotline" between Washington and Moscow during the Cold War. This friendship paid off when we started getting early hints of an RFQ in the making for tactical satellite communications terminals, allowing us to influence the spec to some extent as to what might be practical, and then later perhaps also some "steering" as to what bid price it would take to underbid our competition -- notably Philco, who was a frequent and formidable competitor. This proved to be a very interesting spec to bid, allowing a great deal of freedom as to the resulting system configuration that could be flown in a single C-130 aircraft, with a crew of five, plus fuel enough for several days of operation and with a vehicle for offloading and siting the equipment within a matter of hours. Again, lots of fun working long days and nights on the proposal for what turned out to be our foldable "cloverleaf" concept Mark-V system -- knowingly underbid by a considerable factor in order to win the job. Cost plus a fixed fee, of course. O'Kelley was later able to rationalize somehow that, despite a two-to-one cost overrun, the job had been ultimately profitable for the company because of the large amount of overhead it had absorbed. I was never quite able to grasp that reasoning, but then I guess that's why he was the Chief and I was one of the Indians.

At some point around 1970 Igloo White became a very high priority project in the company, and I was transferred from RF Division to Major Systems to take over the systems engineering job on DART -- I think the folks from Mitre and Hanscom Field weren't satisfied with the guy I replaced (Dave somebody -- can't think of his last name). DART was a quick reaction program (90 days from contract signing to delivery, as I recall) for a "deployable automatic relay terminal" that had an S-band receiving/tracking antenna mounted up on a tower, and a trailer full of electronics ending up with marks made on a paper tape recorder about the width of newsprint, from which trained observers could hopefully recognize patterns of sensor firings indicating VC movements along the Ho Che Minh trail. Sort of a miniature version of the NKP Music Box installation, less the IBM computers. Tom Sheffield was responsible for the paper tape recorder, and Frank Perkins was responsible for a bit rate synchronizer. I spent a couple of weeks up at Duke Field at Eglin with that system during sell-off to Mitre.

Just prior to the DART assignment, I'd spent something less than a year working as more or less staff to John LaCapra, following O'Kelley's decision to fire Jack Robbins and replace him with John. There was never any mutual respect between LaCapra and myself, and once I had helped him update a five year plan for the RF Division (he needed me for that -- I'd been doing it for years for O'Kelley and then later for Robbins), I think he felt the need to get rid of me -- I knew too much and had too many friends.

After DART, I continued in Major Systems running an Air Base Defense Study for RADC, based on some of the same sensor technology involved in Igloo White. I guess this was the period when Lou Goetz, who was in charge of all Igloo White programs, went down to Miami to try to straighten out the common module manufacturing operation, leaving Jack Davis in charge, and when Lou came back Jack was firmly entrenched in his seat. Once the Base Defense study was over, most of us still in Major Systems were pretty much at loose ends as Viet Nam and Igloo White wound down, and the new Harris management couldn't really see much short term profit potential in an organization focusing on future large system pie in the sky, so they abolished Major Systems along about 1971. Between a bad case of male menopause and my disappointment with the new Harris management style (heavy focus on quarterly profits) and their turning away from longer term things that I thought important, I took a three month leave of absence to explore starting up a new business renting motorhomes to tourists flying into the newly-opened Walt Disney World, and then in late 1971 I resigned to actually pursue that dream. Lots of my co-workers were pulling for me to succeed -- everyone seemingly harbored a secret desire to try to break away and make it on their own, and I was actually trying it! Long story short ·the Arab oil embargo of 1973 made what was perhaps a bad idea to begin with even worse, so after exhausting family savings in trying to make a go of the motorhome rental business for three years, in 1974 I finally had to put my tail between my legs and returned to Harris ·this time to Controls Division where I remained for another 14 years, but that? another story.

RECOLLECTIONS OF THE EARLY DAYS
Remarks by ABA at the first Old Timers reunion in 1996

I have particularly keen recollections of the '56 – '57 period when I worked for Bill Dodgson up on the second floor of old Building 2 at the airport. Bill ran what was called "Contracts Department", since you theoretically didn't have to "sell" the government at that time. Our offices opened off a long hallway that had Homer's office at one end, near the head of the stairs, and George Shaw's office at the far end. I can still remember how the whole building seemed to shake as George loped from one end of the hall to the other with his long-legged stride.

Homer had a secretary named Ruth Siebert, who was a steady old gal. Ruth was also the custodian of Central Files, among other responsibilities, and if she found out you had issued a letter or memo without a yellow copy for Central Files, then Buddy you were in big trouble with Ruth! Something had happened to Ruth's voice so she couldn't talk normally – she sort of hissed – but when she was through hissing at you, you were convinced of the importance of a copy for Central Files.

Ruth also thought she was custodian of the Company's morals. I recall that one morning a particularly attractive young secretary reported for work in a sun dress that revealed her shoulders and a whole lot more other territory than Ruth thought was proper, so she threw her own sweater over this girl's shoulders and insisted she wear it until she could go home at lunch and change into something more appropriate.

Homer had an awful lot to do with Radiation's success in the early days – he was far more than just a founder. Among other things he was the Company's best salesman. I remember making a sales trip to Dayton with Homer once – I think Dodgson was trying to make a salesman out of me, and I was to just go along and watch Homer's technique. I think we had submitted a proposal for a telemetry system or something, and we had a meeting with Jim Althouse, who was the Branch Chief of the Telemetry Lab. When Jim walked in, Homer stood up, looked squarely at Jim's top shirt button, and launched into his sales spiel, which went something like this:

"Jim, you know I don't know very much about this except what George and Parker and the others tell me, but they tell me that we have proposed a pretty good system and if you'll just give us the contract then I know you'll be glad you did."

I think we got the contract – but I never did master Homer's technique.

In addition to being a good salesman, Homer was also a very good problem solver. He had keen insight – a way of peeling away detail and getting right down to the heart of things. An example comes to mind. Back in about 1957 we had a large contract with A.C. Sparkplug, and the contract was in trouble. I think we had shipped some or all of the equipment, and they weren't paying us – which was creating a real financial crisis. I think Homer had taken an assignment to talk to A.C. top management about the problem, and meanwhile most of our engineering staff was trying to figure out what was wrong. Shortly after lunch Homer came striding into the Contracts area. (Now you have to understand that striding wasn't Homer's normal gait – he was more an ambler than a strider—so I knew something was up!)

Well, Homer burst into the Contracts area and said, "Where's Dodgson's office?" (Homer never did learn which office was Bill's, even though his own office was only about fifty feet away.) I told him, "This is Bill's office over here, Homer, but he isn't here. They're having a meeting about A.C. Sparkplug down in George Shaw's conference room." Homer said, "Follow me!" and headed down the hall. When he reached the conference room he opened the door and said, "George, could I see you for a minute". George said, "Sure, Homer, we've just been trying to figure out what the problem is with the A.C. Sparkplug system and…" Homer cut him off and said, "George, I've just gotten off the phone, and I can tell you what's wrong with the A.C. system – THE GOD DAMNED EQUIPMENT DOESN'T OPERATE!"

But Homer was really a very tolerant person, and very solicitous. I remember once we had an older engineer named Al Campbell working for us, and the story was that Al had had a very big drinking problem at one time. Homer started picking at Al about that once, and the conversation went something like this:

"Al, I understand you used to drink a lot." "Yes, Homer, I did. " "Well Al, how much did you drink?" "I drank quite a lot, Homer." "Well Al, did you drink as much as a pint a day?" "Hell, Homer, I SPILLED more than that!"

Now Homer would take a drink every now and then too. I remember once we were in Dayton at a Telemetry Conference or something, and Robey Green was flying up to get us in the Piper Apache. He made a stopover in Lexington, Kentucky, for fuel or weather or something, and somehow he taxied into a pile of gravel and ruined one propeller. He called the factory, and they rushed a replacement to Cincinnati, where we were to pick it up and drive it to Lexington. Well, it was a dark and rainy night when we got to Cincinnati, and the propeller just barely fit in the rental car – extending all the way from the windshield to the back window, resting on the seat backs in the middle between Homer and Connie Hoeppner in the back and Bill Dodgson and me in the front. The drive from Cincinnati was maybe a hundred miles on a hilly, twisting road, and Homer had had just enough to drink to profess great concern about Bill running off the rain-slicked road and the propeller flailing around and killing us all, so nothing would do but that Bill had to stop every time Homer spotted an old cardboard box beside the road, and then he'd get out and make a big production out of wrapping soggy cardboard around the propeller so it wouldn't be so sharp when the expected accident happened. The accident never happened.

But I wouldn't want to leave you with the impression that Homer was a person of poor habits or low principles. Quite the contrary. Homer was a man of very high principles, and always tried to stand up for what he believed. I only know of one time when he didn't stand up, and that wasn't his fault. The occasion was back in the '67 or '68 time frame – whenever it was that Radiation acquired Harris's money – and a meeting was being held in the Board Room in Building 2 in Palm Bay to explain the details and benefits of the Harris merger to Radiation's senior staff. I remember that Ralph Johnson, Roland Moseley, and various others were there, and Dick Tullis and George Dively were explaining that no big changes were planned, and that electronics and printing were both communications, etc. It sounded pretty good coming from Tullis, but then Dively, who wasn't as slick as Tullis, said something about Radiation's business being "cyclical". Now most of us didn't have any idea what cyclical meant, but Ralph Johnson thought he knew and started taking issue with Dively, and before long they were in a pretty heated argument. I don't think Homer had a clue about what it meant either, but he could see that Ralph was taking issue with Dively about it, and he knew that he'd spent lots of money sending Ralph to Harvard Business School, so Homer figured it must be something bad, probably disparaging the way Radiation had been run. So he started getting red in the face, and Roland Moseley, who was sitting so he could see Homer better than I could, said it looked to him like Homer tried to stand up at that point to call off the merger – but his billfold got stuck in the back of the chair!

I've probably said more than I should have already, so I'll sit down. But I do want to say, seriously, that I'm proud to have been a part of what we have here.

Ray Buchanan Remembers

I joined Radiation in November of 1963; I was hired by Don Schingler to help install a TAA2 Telemetry Tracking Antenna at Point Magu Naval Air Station in California.

On Friday November 22, 1963, J.B. Williams, Eddy Nelson and I got on Radiation's Aero Commander piloted by Robby and flew to the Grand Bahamas Island. RCA was the contractor for operating the Telemetry site. RCA met us at the airport and drove us to the TAA2 site. We inspected the site and checked the tracking operation of the system.

About noon a radio was on and the news reported that John F Kennedy had been shot. What a strange feeling to be in a foreign country even if it was only 90 miles from the US. Strange thoughts crossed my mind, wondering was this really true. The news reports just repeated the same information over and over again. We left Grand Bahamas about 2 PM arrived back in Melbourne to hear that John F Kenny had died and the rest is history.

The next week I moved from Orlando to Melbourne and readied to move to Oxnard, California and be one of the men to install the TAA 2 antenna. The Navy selected a warehouse that stored atomic weapons to house the electronic equipment. This area was away from the main base and no one came around to see what was going on. We had a QC inspector on site to oversee the installation. Radiation had a crew of about 6 men and about 6 Iron workers on site each day.

Two of the men were J.B. Williams and Joe Young. I don't remember who the others were. Everyday the QC would walk by the fire alarm by the entrance door. He had to touch the alarm lever to see if it would move. This went on for a while. One day about noon we looked outside and there were fire trucks, Naval Police, ambulances the bomb squad on site. The QC guy looked over at me and said doesn't tell them anything. When the Navy asked he told them that it went off on its own. The Navy did rename the fire alarm circuit Telemetry building.

Later on that year Joe Young and I went back to Point Mugu and modified the antenna so that it could receive the Olympics from Japan. NBC had the video rights for the Olympics so they sent the video recording by plane to the US for broadcasting each day. This Olympic video was microwaved across the US then went to Canada and then to a underwater cable to England and Europe. No one in the US saw live video from the Olympic in Japan. In 1964 the bandwidth of the satellites were so narrow only every other scan line was sent over the satellite. The Japanese installed TV equipment would duplicate each scan line of the TV signal, when viewing the TV picture it looked like a normal picture. The Japanese and European TV format has twice as many scan lines as our US TV's.

The last TAA 2 story I have is back at The Telemetry Station on Grand Bahamas Island. Radiation had a contract from Patrick Air Force Base to modify the feed and spar system. Radiation hired a crane and operator to bring it by barge to GBI. After it arrived we flew down to GBI to do the work. As I remember Fred West was the Project Engineer, J.B. Williams, Jim Perkins, Eddy Nelson and myself were on the crew.

The crane had a 120 foot boom and a 40 foot jib. Four of the crew were on the dish, rigged the cable from the crane to the feed and spars. Proceeded to unbolt the four spars from the spar mounts of the dish. When this was finished the four men that were in the antenna dish signaled the crane operator to pick up the feed and spars. The four men stayed in the dish to hold the spars from swinging into the mesh of the dish. The cable tightened and the spars lifted up about 6 to 12 inches above the mounting plates when a cable the supported the jib broke. The jib fell on the feed and the spars fell back into the dish, they fell back on to the spar mounting plates. No one was hurt, the crane company sent repair men over to replace the cable so the jib could be removed from on top of the feed. The jib was lowered to the ground repaired and inspected. The feed and spars were removed safely after the repair. Two days were lost of the schedule was lost due to this mishap. This Trip we flew commercial to the Grand Bahama Island.

Memories of Scott Campbell compiled by Frank Perkins

When an old friend passes it can trigger lots of memories.

Scott came to work at Radiation at about the same time I did, in 1959. We both had previously worked on the same project at Melpar, Inc., in Falls Church, Virginia, although I did not know him there.

At Radiation, we worked in the same Department and were in a group that regularly ate lunch together. A regular topic was technical problems at work, and resolved many issues over lunch. Lunches, with a varied group, continue to this day. At one recent lunch of the current Lunch Bunch, we discussed

Scott's many interests and talents and decided they deserved recording, and I volunteered to compile the results.

First, my memories:

Scott was active in the local Sports Car Club, and participated in many TSD (Time, Speed Distance) rallies. Through his encouragement, I attended many sports car races at Sebring, which was a real adventure in those days.

He had a number of interesting cars. One I remember was a Saab 92, with a 3 cylinder 2 cycle engine, driving the front wheels (a rarity in those days). He also owned one of the first BMW's in the area, and several over the years.

Scott contributed greatly to a number of programs at Radiation, including an advanced packaging scheme using molded modules. but he was not fully appreciated and eventually left to become an independent engineering designer, working on many projects, particularly advanced welding systems.

I talked to Scott's first supervisor at Radiation, Al Johnson, expecting him to have many stories of the difficulties of channeling his efforts. Al only remarked that Scott was difficult to keep track of, because of his many self-assigned projects.

Scott had many and varied interests. He was proud of his Scotch heritage and was active in the local Scotch Country Dancing Club.

John MacNeill, one of Scott's oldest friends, responded to my request for some memories with the following:

The first of Scott Campbell's characteristics which comes to mind is the thoroughness with which he researched all purchases and choices. One example was finding the right material to use for some flexible tubing that could not lose pressure over an extended period of time. His investigation turned up the fact that Saran was the best available.

The same applied to his selection of automobiles. During a test drive (on the dealer's lot) he alarmed the onboard sales person while checking out the vehicle's handling capabilities in close quarters. Since nothing got hit in the process he concluded that they were pretty good.

He was very interested in sports car racing and after a race he and Ann (Scott's first wife) met socially with the Marchal's (famous maker of high-powered automobile lighting). He knew personally the Porsche importer in Jacksonville. I went with him on one of his tech inspection assignments and he didn't like the looks of the driver's seat of one car. I did a simple analysis of its strength and found that Scott was right and that the seat was way under strength. The owners reinforced the seat in time for the race.

On the technical side, in addition to the Metal Cladding Projects for Dick van Loonan, Scott worked on an IBM card reader with Hasty Miller and started to develop a tape punch, and later an impact type dot matrix printer. Our interests were so nearly the same that a postman who handled both my mail and Scott's commented on the similarity of the catalogs we received for technical equipment.

Other projects included a device to dispense shots of liquor precisely and an improvement on an ultrasonic cleaner, the latter project involving a large piezo-electric disc which had to be securely bonded to the bottom of a small stainless steel tank - another research project to find the right adhesive. His client marketed these cleaners through Sears stores where they were sold as a cleaner for false teeth at $30. Sears did not realize that they probably could have sold more in the tool department for cleaning parts. I bought two for the shops for cleaning automobile parts and when I checked one recently, it still worked after several decades. This was determined by putting aluminum foil in the liquid and observing that holes were punched through the aluminum by the cavitation caused by the shock waves in the tank.

Other items of interest to Scott were classic music for the enjoyment of which he bought and made top quality audio equipment. Food was something he knew a lot about. Bread pudding was one of his favorites. He knew what went into "the haggis" but ate it anyway. Scott accompanied Ann on incognito visits to local restaurants, inspecting quality of service and food.

Scott's interest in his Scottish roots was such that he wore the kilt and sporran on occasion, which was often. At one of these events, his mother and I were watching him busily doing his duty for one of his Scottish groups and she said, "I don't know what got into Scott; he was such a studious boy." The same interest took him with Ann on a trip to Scotland and in their usual style they made close friends with the Duke of Argyll, who was the chief of Clan Campbell.

The university Scott attended required that electrical/electronic engineers take many courses normally of interest to mechanical engineers. Before World War II I had graduated in Mechanical Engineering, but came back to study electronics and servo theory. Scott and I jointly applied this training and our experience to judge Florida State Science and Engineering Fairs for decades. Due to his outgoing nature he was often his team captain.

Another capability uncommon among electronic engineers was knowing how to use the milling machine/lathe he had in his rather extensive machine shop, which was separate from his electronic shop. For example, he made bronzed rocker arm bushings for his BMW automobile, which lasted longer than the factory product. The research here was to determine the configuration of the oil grooves to be cut by hand on the inner surface of the bushings.

Part of Scott Campbell's legacy is the "Show and Tell" group that has met for lunch on Fridays for about four decades. He put together a group of six persons, the members of which had such diverse and common interests that conversations started during lunch are finished on the sidewalks outside. The subjects covered are individuals' projects, automobiles, airplanes, and boats; there are no off-color jokes or raucous laughter, and very little politics and religion, but many amusing experiences and a few puns. (John's wife Maggie reports many late night phone calls from a lonesome Scott (after Ann died) covering an amazing set of topics. Hopefully she will be able to supply some sort of compilation.)

Tom Bowman, of the Lunch Bunch, supplied the following:

Unlike some of the others whose memories of Scott are recorded here, I never had the pleasure of working with Scott, although some of my former students in the mechanical engineering department at F.I.T. ended up working at Harris and Scott commented to me more than once on what good engineers at least a couple of them were. Also, Billy Thrasher and Frank Stasa from Harris, who started teaching some of our courses in the 1970s and ended up joining the M.E. department at F.I.T. while I was its Head, both knew Scott at least by reputation and possibly from personal experience as well, so I can't help feeling I knew Scott at least obliquely for more than 35 years.

We first met in person when John MacNeill (whom I first met at a local ASME meeting in 1969!) invited me to join him and Scott in judging the State Science Fair, which was in Fort Pierce that year (maybe 1975 or 76?). Scott drove us there in his wonderful BMW 2800, and even put his foot down a bit coming home out of Vero Beach, so it was a memorable ride in a most impressive and comfortable motorcar. As I recall, Carmen Palermo from Harris joined John and me in judging the ME projects while Scott was busy with the ones in EE.

(I was never able to join them again because my own 4 kids got involved in Science Fair over the next umpteen years and at least a couple of them went on to the State Science Fair and even the International Fair, so that put an end to my fun of being a judge until more recently and by then it seems most of the fun was gone anyway, for me, what with most of the projects dealing now with pretty esoteric high tech stuff. One of my bosses at F.I.T. always told me I was just disgruntled because steam-driven computers never caught on!)

In the years that followed I came to realize more and more just what a multi-dimensional man Scott was. One example that comes to mind was a musical performance by a group of young Finnish singers that was touring the U.S., probably some time in the 1980s. Ritva and I bought tickets (since she is a Finn) and when we got to the church where they were preparing to perform one of the first people I saw was Scott, standing in the back coordinating many aspects of the preparations and organization, and at the same time finding the time to talk with me on various topics, some related to the event and some not.

I don't recall any other specific instances, but there must have been several because it seems like I have always known about Scott's SAABs and his extensive racing, officiating and scrutineering activities, and because of our many common friends and acquaintances where the link was a mutual interest in small sporty cars from small British and Swedish builders.

Some time in the second half of the 1990s we ran into each other in Home Depot after having lost contact for several years. We recognized each other instantly, although Scott was already having trouble with his balance and his walking, but his mind was remarkably clear and his quiet enthusiasm was obvious, and we must have stood in that aisle and talked for more than a half hour plowing over fond memories from earlier times and filling in some of the blanks from more recent years. Before we finally moved on Scott asked me to join the Friday group that gets together each week to share thoughts about all kinds of techie topics and issues, some current events, some interesting new historical discoveries, sailing, cars, theories old and new, almost anything except politics and religion. Come to think of it even some politics reared its ugly head during the time when the right wing was stealing the presidency away from the man and party who got the most votes—maybe the reason Scott wanted me to join the group was so he would no longer be the lone voice of reason trying to bring the others to their senses!

Scott, I hope you're getting all this!

Hasty Miller reports further on Scott's automotive skills:
I first met Scott at John MacNeill's house. Scott was a friend of John's and our mutual interest was cars. In 1970 I wanted to try to make a living as a consultant, and Scott had a client who wanted a card reader designed.

So I quit Soroban and went to work with Scott on a new type of card reader.

That project went on for a year, and I continued to consult for a year after that. During this time, Scott and I traveled to South Florida many times in his new BMW 2800. He demonstrated to me one time what "Cruise at 100 MPH" meant. We were driving comfortably along listening to the stereo while the telephone poles passed like pickets on a fence. On our trips, he took me to several good restaurants that I had never heard of. One of his hobbies was finding good restaurants.

On one of these trips, we were in Ft. Lauderdale, and we saw a '54 Studebaker Coupe sitting in a gas station with a for sale sign on it. At my urging, he pulled in, and we inspected the car. It seemed to be in good shape, so I asked the price. The station owner said that the owner wanted $150 for it. $50 would hold it until I could get back down with the rest of the money (this was before ATM's). So Scott and I pooled our lunch money, and the next day I came down on the bus to pick up the car, which I still have.

I contacted the Florida chapter of the Sports Car Club of America, and Bill Martin responded with this email:

Yes, I remember him now. He was a close friend of Joe Oliver who was the Chief of Tech Inspection at that time. I think Joe also lived over on the east coast. I know he was active in Tech Inspection around '69, '70, '71. Sieck has been compiling Regional records. Perhaps he can find something. ---bill

Another member of the Lunch Bunch, John Butler, reports:
Scott was a member of the Board of Directors of the South Brevard Sharing Center, a not-profit agency that provides food, clothes and emergency funds for people in need.

Scott many interests made him a great conversationalist. Lunch is not the same without him!

Frank Perkins August, 2007

Bill Corry Remembers

A talk to Radiation Old Timers April 16, 2011.
Maybe it should be entitled a talk to the Alzheimer's Association. This was the start of the joke, but I forgot the rest.

I was hired by Radiation as a detailed draftsman for the summer of 1953 and if I remember correctly, my employee number was either 113 or 118. I worked for a man we called "Whip Wilson". I became proficient in running the Ozalid machine and actually enjoyed the ammonia smell. I did not know it at the time, but I was considered a co-op employee while I was going to the University of Florida. The drafting room of about 6 employees was located across the street from what is now the main terminal of Melbourne International Airport in Radiation Building #2. The building *was* left over from WW2 but served the purpose for our executive offices (Homer was on the Second floor on the East end), our machine shop (Headed up by Wally Silver and managed by Leo Hosley), some engineering offices and the drafting room. I seem to remember creating drawings for project number 1003 which was an Underwater Towed Sonar. If memory serves, the experimental test of the Sonar was conducted in the Indian River from Homer Denius's boat. I suspect Ralph Johnson or some of the engineers of that time could give a good explanation of what happened when the towed sonar was lost in the Indian River. (As a Program Manager many years later in my career, I was given the task of completing the process of closing out the Towed Sonar Contract that had been overlooked by accounting and finance earlier).

My second summer exposed me to the details of what became the Teledeltos Recorder, a concept of writing on sensitive paper with an electric arc at high speed. There were several early projects that used this concept of writing. Each writing element (of which there were maybe 128 or so) required a vacuum tube driver that generated considerable heat. I worked with Jessie Ferguson during this summer and with Carl Bromm. Bu Shing Chen and (I think) Gantt Hamner were Mechanical Engineers during this time. I was grateful for my Drafting job *as* the $1.35 an hour salary the first summer helped put me through college.

I do remember a brief period when I helped our company pilot, Robie Greene, dispense chaff from an airplane over the Valkaria Airport. We were making some radar measurements and there was some problem with Florida Power and Light claiming we were shorting out some of their power lines. Robie was a character and while he flew the Aero Commander and our LearJet for a time, he also played the clown in a Piper Cub demonstration at the Melbourne Airport. He made it look like he was drunk (maybe he was) but it was a fun series of acrobatic maneuvers.

After graduation, Personnel Director Ralph Bickford offered me a job for a little over $325 per month working with Don Gallentine on various projects, my first being to design a shaft position transducer that used strain gauges tracking a cam shaft to report angle. Not too exciting but a big deal for a new engineer. After only 8 months, I was having a difficult time taking orders and at the suggestion of my father, went to work for Bell Aircraft in Buffalo, NY testing various rocket motors. After only 3 months at Bell, George Shaw gave me a call asking me to return to the Radiation facility in Orlando. I negotiated a small raise and noticing 4 feet of snow outside in early December, I made it back to Melbourne for Christmas. The New York authorities hounded me for years to pay the taxes on my earnings at Bell Aircraft. I believe I am beyond the Statute of Limitations.

This time, I remained with Radiation and later Harris for an additional 31 years before doing some consulting and then returning for a final time with a total accumulation of 37 years.

While in Orlando at what is now known as the Orlando Executive Airport, I worked on an AVCO nose cone for the Atlas Missile. We instrumented the Nose Cone which was to be dropped from an B52 to learn its flight characteristics. It was a successful project. Julian Zenitz was my mentor. Now that I was properly employed, I married Evelyn before she ran away to Flight Attendant School. This was over 54 years ago.

After maybe a year, the entire Instrumentation Division was moved to an abandoned laundry facility on Boone Street in Orlando. Here, we undertook to find out what happened to the instruments on Operation TeaPot, the Atomic Weapons Test near Bikini Atoll. We shook the equipment, placed it in a shock wave chamber, centrifuged it, and did drop tests to simulate the environment the equipment experienced during the bomb blast. By experimentation, we learned that best way to simulate the force profile of the blast was to drop the equipment on toilet paper on edge. We used a lot of toilet paper that summer. This was about the time when Parker Painter moved out and started his own company in competition with Radiation. The Boone Street facility was just temporary until the new plant was constructed just east of McCoy AFB, which was a SAC base at the time.

I seem to recall that the Instrumentation Division moved into the new McCoy plant after about a year. I worked for Ralph Parr with Bob Dimond and with John Mazur. Just did a bit of packaging of the Teledeltos Recorder equipment I mentioned earlier.

Soon after, I returned to Melbourne (the dates are a bit confused). I think I worked first for Bob Marquand assembling a Rocket Sled Instrumentation Package for White Sands and developing the Radicon/Radiplex commercial PCM equipment. This was the time when we bid and won the Minuteman PCM Telemetry under subcontract to Boeing. I worked with Bill Eddins to package the missile borne equipment and to write the proposal. This was the largest contract that Radiation had won up to this time. I think that the nearly one cubic feet of electronics that we built then could be replaced today by a single computer chip.

In Palm Bay, I worked for Harvey Bush while we went through various missile and satellite programs such as Telstar, AC Spark Plug and Nimbus. Harvey was good enough to allow me to tackle some of the programs involving structural and mechanism designs. We developed the Telscom Antenna System with LaVerne Williams and had a short run designing an antenna for the ill-fated DynaSoar Program. We then began getting into various large antenna projects known as Bayhouse and Stonehouse to be located in Ethiopia. As a frame of reference, we redesigned an NRL 150-foot diameter, 130 ton, steerable antenna, built it in the USA, shipped it, and installed it in Ethiopia in 11 months for just over 1.1 million dollars. Then there was the Mark V portable communications antenna design headed up by Jack Johnson which had to survive the Munsen Road Course in the parking lot in Palm Bay. Later, working with Dick Baker, we bid the design and construction of the Altair 150-ft diameter antenna in

Kwajalein. Harold O'Kelley made an urgent appeal to the head of DARPA in the Pentagon and with Cole Goatley's sales finesse, we won.

But now I am starting to talk of projects during the Harris Corporation era and this is a Radiation's Old Timers talk. I'm done.

MEMORIES OF MEL COX by A.B. Amis

Mel and I were friends for a long time -- more than half a century, actually. We were classmates and lab partners at Georgia Tech, and finished together in 1951. We interviewed several of the same companies, and both ended up accepting our highest-paying job offer -- from a small electronics company in Indiana -- which was probably a mistake. After one very hot summer and one very cold winter there, we were both ready for a change. I had spotted an ad for another small electronics company -- this one in a place I'd never heard of called Melbourne, Florida. But Mel knew where it was -- he said it was close to where he and Joyce were heading for the Christmas holidays anyhow, so I corresponded with the company and made arrangements for Mel to stop by and look them over. He actually ended up interviewing for both of us while he was there, and they hired both of us on the strength of that one interview. I don't know if I ever really thanked him for getting me the job, but I can confess now that I've maybe always carried a little bit of a grudge that he managed to get a higher salary offer for himself than for me. But that may not have been any of his doing -- the company may have just given more weight to his Navy officer training than to my better grades.

Mel was smart -- probably smarter than me, in spite of poorer grades in school. We approached college, and I expect many other things in life, differently. I guess college was like a game with me, and I won if I made good grades on tests, even if I never developed a real understanding of the subject matter. Mel, on the other hand, had far more actual interest in and curiosity about the subject matter, and taking tests and getting lab reports in on time were just bothers to him. If a particular subject or aspect fascinated him, he might just dawdle there while the professor and the rest of the class went on to other matters. Those of you who worked with Mel know that he carried these habits on over into his professional life. If he came across some new scientific or mathematical oddity that he hadn't been aware of before, then he was sidetracked for whatever time it took him to get his fill of studying or playing with the thing.

I mentioned earlier that Mel had received training as a Naval Officer. He came away from that with some firm ideas about chain of command and the importance of authority being delegated along with responsibility. These ideas kept him in a state of pretty much constant turmoil with his managers early in his career. The failure of his bosses to meet his idealized expectations of what bosses should be frustrated Mel and frustrated his early goal of a career path in line management. A pipe-smoking misfit from Kellogg named Dick Hultberg comes particularly to mind. Mel summed up his frustrations with the process by which managers are chosen, by developing a critical tongue-in-cheek model of the process, which he named the "Blooper Bag". I'm sure that everyone here who worked with Mel had that model explained to them at one time or another, just as they had corner reflectors and other such technical oddities explained to them. I don't recall hearing as much about Mel's Blooper Bag in later years – perhaps he was mellowing a bit, or more likely it was that his later staff assignments in program development didn't place him in conflict situations with his management.

Mel was a good salesman and a good recruiter. He was a quick study – grasping key elements of a technical problem or a marketing situation in a hurry – and he was a good writer. But there was something else that seemed key to lots of his successes. He seemingly had an ability – and I don't know how he did this – to inspire confidence and loyalty in a customer. Customers somehow trusted Mel to see to it that their project got a good effort from Radiation, or later Harris. Most of you know the key role that Mel played in securing early contracts placing Radiation/Harris solidly on the ground floor in satellite communications and surveillance – leading to billions of dollars in subsequent contract awards that continue to this day.

I mentioned that Mel was a good recruiter. Let me recite one anecdote about that. This happened early on, when Mel was in charge of a radar reflectivity facility (which, incidentally, generated good "bread and butter" work for the Company year after year – based largely on the strength of the relationship Mel maintained with the customer's project engineer). As I recall the story, Mel was interviewing a bright new college graduate, and the interviewee asked what he'd be doing if he took the job. Mel reached into the bookcase behind him and removed his copy of "Reference Data for Radio Engineers", thumbed his way to page 467, and pointed to an equation about 2/3 of the way down the page. He then said "This is just one book out of that bookcase full of books, and one page out of something over 600 pages in this book, and one of the two equations on this page. And if you look at it you'll see it's the radar range equation, which has eight or ten terms in it – one of which is this term, sigma, which is a measure of the radar reflectivity of an object. Well, I've made my whole career up to this point trying to measure and learn about this one term in this one equation on this one page of this one book." I don't know whether that young man took the job or not – but lots of people did. Incidentally, when you see TV images of those new weird angle-shaped stealth aircraft that can avoid radar detection, remember that story about Mel – that was an objective of the work he was doing forty-some years ago – trying to hide aircraft from radars.

But Mel wasn't all business – he had a whimsical streak in him too. That streak was pretty wide, actually. He had a particular affection for nonsense words and names. Pfleugerhaven comes to mind – that's what he and his son Jamie labeled a goofy two-wheeled contraption they built together. They were thirty years ahead of their time – There's a new gadget called Segway employing somewhat the same principle that's now being touted as something that's going to revolutionize transportation.

Let's see what other silly stuff Mel did. Well, one of his prouder accomplishments was being able to touch the tip of his nose, or his chin, with his tongue – and having the dexterity to tie a knot in a cherry stem with his tongue. He knew the words to an awful lot of smutty songs he'd learned in the Navy, which served us well at beach parties when we were all younger and wilder. He could recite fifteen or twenty verses of a poem about a farmer's dog named Runt, and he could spot the planet Venus in the daytime -- it wasn't uncommon at all to see Mel in the middle of a group of people in the parking lot after work, pointing into a seemingly empty sky. He wasn't above pulling a prank on one of his co-workers – but of course that worked both ways, and I expect Mel actually got the worst of that on balance, because he also tended to be gullible. I can remember his telling me that when he was a little guy he was being raised in a house full of women – his Mom and maybe some aunts – and they all took delight in making up fanciful stories about the strange people they'd encountered on the way home from work – just to see Mel's eyes widen in wonderment!

Mel used to enjoy fishing, but other than that he never was very big on the outdoors or exercise – he usually preferred working on some project in his "shop" – whether it be ham radio or woodworking or computer genealogy. But in spite of very little exercise, he never seemed to gain weight. Part of that had to do with his heavy smoking, I expect. But I think he also had a very unusual metabolism that allowed him to eat three or four hot dogs at a sitting, and instead of converting that to fat like the rest of us might be prone to do, he converted it to acid sweat that would rust out the top drawer of a new steel desk in about two years, and turn a set of brass car keys into ash in a not much longer time. He used to enjoy playing a Jews harp, until the "twanger" rusted off.

Mel actually played that Jews harp with more enthusiasm than talent, because he really didn't have a whole lot of musical ability – but he did have an appreciation for music – particularly big band music from the 30s and 40s. At some point in his young life he'd worked late nights as an engineer at a radio station with a huge library of big band records, and in his later years Mel took lots of pleasure in downloading some of these old selections and burning CDs for friends.

Sailing was an important part of Mel's life after retirement, and he even got into "the arts" briefly after getting hold of equipment for forming and firing ceramics. Mel took particular pride in a few ceramic sundials he made and gave to a handful of his friends. His pride wasn't so much in the aesthetics of the sundials as in their mathematical perfection. He'd gone to great lengths to see to it that the length and angular positioning of the "gnomon" (now that's the kind of word he loved – it looks and sounds funny, and is something most people never heard of -- it's the little stick that projects up and casts the shadow) were exactly right for our latitude and longitude here in the Melbourne area. It didn't really make any difference in a practical sense – it was just the satisfaction of knowing that it was mathematically sound that pleased him. He had let that sundial work as it was supposed to work.

Mel enjoyed such mathematical challenges. That's why I turned to him for help when I needed to design a wooden barn – I knew he'd remember far more than I about calculating moments and stresses and creating a geometric design best able to resist the wind forces dictated by the building codes. He dragged out old college texts that he'd never thrown away and set to work on this new challenge, and working together we came up with a design that delighted him. He insisted on being there when the crane came to set the massive cypress posts and beams into position and the design began to take physical form, and for years afterward when he'd come down to visit he'd say, "A.B., let's go out and look at that magnificent barn!" He reveled in the strong geometry created by the sturdy posts and beams and angled braces working together to provide rock-solid strength.

One more story about Mel. Once on a business trip out west, we found ourselves with a couple of hours to kill before a flight out of San Francisco, so we went to a bar a few miles south of the airport for a beer or two. We were sitting at the bar, right in front of the beer taps, and Mel got to admiring one of the tap handles that was particularly ornate and whispered to me that he was going to steal that handle. So a few minutes later, when the bartender's back was turned, Mel gave the handle a couple of turns counterclockwise. The next time the bartender went to draw beer with that tap the handle sort of wiggled in his hand and he looked at it funny. Noticing the problem that he seemed to be having with the handle, Mel commented on it and said, "Here, let me have a look at that handle – I'm pretty good at fixing things, and there seems to be something wrong with it." So he gave it a couple more turns counterclockwise and asked the bartender to try it again. Well, it wobbled even more, so Mel said, "That's what I was afraid of – it looks like it's broken and you'll have to get it replaced. That's too bad, because it's a mighty pretty handle. I wouldn't mind having one like that – how about giving it to me, since you're going to have to replace it anyhow?" The bartender said it wasn't his to give away, that only the boss could do that. So Mel asked him where he might find the boss, and the bartender pointed him out, sitting at a table over on the other side of the room. Mel got up and went over and introduced himself to the bar owner and sat down at the table with him for a few minutes, making small talk, and then returned to the bar and said, "Frank said it would be OK for you to let me have it" and finished unscrewing it, put it in his pocket, and we left.

Mel died in early 2002 following several years of declining health – precipitated probably by the heavy smoking that he was seemingly never able to give up for long. I came to the realization in the few days following Mel's death that in the fifty plus years I'd known Mel, we never had many conversations about "soft" subjects like philosophy or religion, and how he "felt" about lots of important things like life and death. I guess we somehow always talked about whatever Mel wanted to talk about – either problem solving or else the latest mathematical oddity that had struck his fancy. But maybe I can provide one final insight by reciting another anecdote that happened while we were still at Georgia Tech. A professor, and I believe his name was McKinley, was lecturing on modulation of radio frequency waves, and he introduced what Mel considered to be a hocus pocus concept of "sidebands" as a means of visualizing and understanding the modulation process. Mel raised his hand and asked if the professor really believed these sidebands existed, to which the prof replied, "Yes I do, and you will too if you expect to pass this course!" But Mel never did buy into that concept of sidebands – he placed his faith instead in the mathematics that underlay the sidebands. He just never brought up the subject again. And he passed the course. Mel believed in the unvarying principles and constants that underlie mechanisms and systems and institutions that work well – and work as they should – I believe that's the temple where Mel worshipped.

Don Sorchych remembers Urie Davisohn

A woman doing research and writing about Uryon Davidsohn recently contacted me asking about a brief reference I made about him in an August 2008 My View column. The original publication date of this column was February of 2008.

In 1965, five years into my career after graduating from the University of Illinois with a BSEE, I was faced with a dilemma. A couple of years prior, I had been offered a substantial promotion to move to Silicon Valley and join a semiconductor company. The little company I worked for, Radiation Inc., had flirted with starting such a business and promised I would be able to transfer there if I stayed.

Radiation, after a search, hired Uryon Davidsohn, who had been a freelance staff scientist at General Electric. He started the Physical Electronics Division charged with the goal of developing integrated circuits.

Underfunded from the start, nothing much materialized so when I was offered the Director of Engineering position, I refused. Pressure was applied and I reluctantly agreed to take the position.

Although "Urie" was still the head of the division, corporate had sent over a V.P. of marketing, Jack Hartley, to find out why so much money was being spent with no tangible outcome.

Urie was a short, humpbacked, bug-eyed, homely guy, exceedingly brilliant, weird, but with a certain charm.

Urie's staff meetings were a riot, with a gaggle of engineers mostly hired from Texas Instruments. These engineers were out of control, undermining each other and trying to figure out why the limited output of circuits suffered from a failure mode no one understood.

A technician named Ed Guerra, also from Texas Instruments, found the answer in a technical paper published by IBM, but what did a technician know?

Eventually, Guerra's findings solved the problem.

Urie had unusual ideas and decided to build his own photo masking method rather than using industry standard step and repeat cameras. He ordered a precision Plexiglas board with holes drilled at regular intervals to hold film. It took a year to have it produced. He placed an array of high intensity lights behind the board to illuminate the film.

When it arrived and the lamps were turned on behind the eight by eight feet Plexiglas board, it began to melt.

That was the end of that project.

Tension between Hartley and Urie mounted, so Urie went to corporate management with a series of allegations about Hartley, which were untrue, and raised the tempo of animosity.

Somewhere along the line, Urie divorced and bought a bachelor pad on Merritt Island. He was an extremely anal person and a clean freak. He invited Hartley to his home. It had been raining and he told

Hartley to clean his shoes on an outside carpet.

When Hartley came inside, Urie pulled a four barrel .22 caliber Derringer from his pocket and shoved it in his face. That resulted in Urie being involuntarily committed.

When he was released, he would call me at all hours of the night and threaten suicide unless I came and had a brandy with him.

He had a collection of antique stringed instruments and he would sing folk songs while plucking strings. It was agonizing to listen to.

One morning at about 3 a.m. he called and said he was with his ex-wife who he was going to kill and then commit suicide. I raced up there only to find him singing love songs to her, but with his ever present Derringer at his side.

Finally, he began a long soliloquy about what a decent husband he had been, never sparing money, being a loving, caring father to his children and on and on.

Then he told his wife to bring brandies for us. She handed each of us a drink and sat down. He looked at his glass, looked at mine and yelled, "Bring me Don's glass."

She did.

He held the glasses up side by side and screamed, "You f***king bitch you gave him more than me!"

Later he decided it was important for a teen-age daughter to experience sex, but under his supervision. He took her to the Bahamas in his small boat. His boat experiences are another whole series of stories.

He followed her from club to club until she was picked up by an Irish seaman. He followed them down to the wharf, and when they got to the boat, the seaman turned around and said, "Get lost old man."

Out came the Derringer. Two quick shots near the toes of the seaman led to a quick retreat, as fast as the seaman could run.

Urie's secretary was a tall, attractive woman named Laura. She adored Urie, called him Dr. Davidsohn and held him in high regard, so she was pleased when he invited her to dinner at the Melbourne Beach Steakhouse.

When they got in her car, he reached over under her dress. "Dr. Davidsohn, what are you doing?" she screamed while pushing him away.

He said, "Well, I bought your dinner, didn't I?"

Laura said, "How much was it?" He told her. She reached in her purse, counted out the amount and gave it to him.

Somewhere along the line an investigation revealed Urie was a phony. He didn't have a Ph.D. as claimed, among other things. Although he held a high scientific position at G.E. it was not clear whether his brilliance stood on its own or whether they were fooled too.

Urie joined his friend Dr. Arnie Lesk at Motorola in Phoenix. Lesk was in charge of a research lab. I followed Urie's career through technical papers, usually about a method called dielectric isolation.

I think he remarried and two of his sons were either visiting or living with him.

One day Urie told one of his sons to clean the pool. He resisted. Urie insisted, apparently strongly.

The boy went in the bedroom, retrieved a shotgun, came out and blew Urie away.

Jack Hartley was promoted to another division and later became President and CEO, then Chairman of Harris Corporation, who acquired Radiation in 1967.

I took over the Physical Electronics Division, which later became Harris Semiconductor.

R.I.P. Urie.

Homer Denius Remembers

The first time I came to Florida, went to the Patrick Air Base to watch them shoot a V-2 rocket. I had never been out in this type of country before. It was all palmetto and sand. They had built a small road to the lighthouse area of Canaveral to take the missile in; they had tried a couple of shots before, and all had been failures. Werner von Braun was there, with the commander of the base, several military people, the lighthouse operator, and a few others. They touched the thing off and it went right on up. It was the first successful shot of the V-2 in the United States. Everybody was very happy. The people working there had gathered in a bunch of rattlesnakes that day and had stretched them out - they were very large.

The fact that we were able to get in the digital space technology business before anyone else, and did a very good job in that respect, was largely responsible for the growth of Radiation. We didn't have much competition and our group had more knowledge from the start in digital telemetry than any other group in the country. I don't think we ever lost a contract in that particular area.

We had a sales manager who was afraid of bears. One day, George Shaw and I were in a car headed for the plant and we came across a hitchhiker along the road. He had been with a carnival and had a trained bear with him. George told the man to get in, and bring the bear. When we got to the plant, the sales manager was out to lunch and George got the idea to hire the man and the bear for an hour or so. When the sales manager came back and opened his office door he found the bear sitting at his desk. There was quite a bit of commotion for a while.

One big lesson I learned when I was quite young, and I've always followed it. Never make a promise you can't keep. And it's really paid off. It doesn't matter, as far as I'm concerned, whether it's business, or friendship, or something else. You should never, never make a promise if you can't keep it. That's the way I feel about things.

When I was in the fifth grade in southern Ohio I made my first radio set out of a cardboard oatmeal box and a crystal. Somebody gave me a set of headphones and some wire and I could get KDKA in Pittsburgh. Then another fellow came along and gave me a lot of radio parts — a three-tube set and a whole bunch of parts to make radios out of. That's what got me started, and I never got away from it.

Bill Dority Remembers

This could be fun. Let me recall to you my early days with Radiation. I came there in the Spring of `65, working for Ralph Johnson who headed up a "Controls" Division located in the Barnes Engineering Bldg out close to the Airport. They had just recently completed the acquisition of the controls products of North Electric (remote control and Locotrol). Now someone in Radiations management had convinced themselves that you could not be a member of the ELECTRONICS ERA unless you had a computer product line. Since I had just been let go by a company who nearly went bottoms-up following that same drum beat, my job was to put Radiation into the Computer business. The first week I was there, I joined Ralph and a couple of Engineers from the " Data " group at Palm Bay to go to New York at an Electronics Show to evaluate a main frame computer being touted by Union Carbide Corp, who was also trying to diversify into computers. It turn out to be a shell with a lot of blinking lights. Fortunately, we walked away from it. Still the passion existed there for a "computer" of our own,. But, when it came time for me to put together a Business Plan for going forward into the next Fiscal Year, I was stumped. All the data said it was a waste of time and money as proved by the published list of companies that had tried the previous 2 years and had gotten their butts burned. So, my recommendation was grow and expand the Controls Product lines, using other peoples` computer hardware and concentrate on being a Systems Supplier to the electrical, pipeline and railroad entries provided by North Electric. That was not a very happy moment in my career at Radiation, let me tell you.

However, shortly thereafter, they folded Johnson`s group into what became the Controls Division, canned him, and offered me a marketing job at Palm Bay. I had a choice between RF, Aerospace and Data. I chose Data, again to the consternation of quite a few, and went to work for Rollo Christenson. Boy, it was rocky ground there for the next year until my team won the SGLS Ground Station Decom contract from the Air Force in Los Angeles. Lots of Hoorah for that, only to wake-up later and find that we were 45 days late on getting the contract staffed and underway and the AF was starting to build a show cause action against the company. What had happened is that the Vietnam War created a market for Electronic Warfare products to try to root-out the Viet Cong. The head of the group at AFESD was Lt Col Dildine who while at AFSSD led the selection of us for the SGLS contract and came in with a lot of RDTE cost plus contracts to Radiation that gobbled up the management attention and resources from the SGLS contract. End result, while we were heroes in the eyes of ESD we were lower than low in the eyes of the group in Los Angeles. It wound up with Jack Hartley having to go out there with hat in hand and on bended knee pledge the company jewels to prevent cancellation of the contract for cause. In the end, a very excellent product was delivered with the company eating over a million in overrun on a 2.2million FPI contract...`For several years afterward whenever I presented a program at the bid no bid and /or the pricing meeting, I had to withstand the query from Accounting of " another SGLS job, Bill, or why can't we get more of that Vietnam type job instead of this risky business?" And then, "what is this 16 kbps modem foolishness? RADC never comes through with anything. You're wasting our time." The hardest part was selling within, right!. But it was fun, to say the least, and I enjoyed every moment of it.

Jay Fleming Remembers

I was hired in 1957 and my first assignment was at the Malabar site, where we were measuring the characteristics of "chaff" used to mask aircraft from radar. Chaff was metallic strips whose length was tuned to the frequency of the radar being jammed. The long lengths required for relatively low frequencies (we measured from X Band to 250 MHz) were a hazard to power lines, so the tests needed to be carried out in a remote area, like the outback of central Florida. Radiation had a reputation with the Air Force for high quality chaff measurements, so we enjoyed a series of contracts.

One of the tasks for our group was to fly a balloon with a reflector to calibrate the test radars. This involved driving to a remote area and waiting while the measurements were performed. This could be boring and lonesome, and it wasn't a popular duty, so everyone was surprised when I volunteered, but I had an ulterior motive. I had noted some ponds and water-filled ditches, so I took my fly rod and had a good time catching fish while the tests droned on.

Another entertainment while flying the balloon was visiting the "Anderson dig", where A.T. Anderson, an eccentric ex-tool and die maker had found ancient animal and human bones. Anderson's only drinking water was sulfur water from a deep well, a vial tasting and smelling liquid. When I asked for a drink, he suggested doctoring it up with a little orange syrup to make it more palatable. It so happened that my boss came by about this time to see how I was doing. I asked if he wanted a drink. He looked at me and the whiskey-colored drink in a funny way. "Let me smell that," he said. One sniff convinced him that it was harmless, if not very tasty.

Another program I remember was a Doppler navigation System for Sperry Gyroscope. This involved lots of circuits and devices we only barely understood, but we finally got a prototype working. For testing, we leased an old Beechcraft, tired veteran of navigator and bombardier training in World War II. It seemed to take an inordinately long time to leave the ground on takeoff. The pilot, not very experienced in the type, mentioned this to another pilot one day. "What you need to do," he was told, "is pull the gear retraction lever up during the takeoff run. They won't come up until the weight goes off the linkage, but then the reduced drag lets you get up to climbing speed faster." It sounded like a plan to our pilot, and he implemented it on the next takeoff, with a cabin full of test personnel. Unfortunately, one of the things worn out on the plane was the linkage latching the gear down, and it dutifully retracted on command, dropping the plane to the runway and bringing it to screeching halt. It took a while to repair the equipment and the confidence of the engineers in pilots and aircraft.

When the equipment was finally delivered to the customer, The Radiation delivery engineer encountered incredible level of union and company rules and regulations in regard to moving and setting up the equipment. Finally, after racking up 28 rule violations in one day, he came home, to the mutual relief of all parties.

People I worked with on the Doppler navigator include Jack Harrison, Houston Singletary and Edna Tutwiler.

Memories of Bernard France Collected by Frank Perkins

The whole France family was close to me, largely through my long association with Bernard. Bernard and I ate lunch together for over 40 years, so we shared lots of memories and associations.

We came to work at Radiation at almost the same time, in 1959. The first job we worked on together was the Boeing Tape Format Converter, where our Project Engineer was the delightful Larry Klingler. Larry was very exacting in his design approach, particularly insisting on logical, intuitive control panels. I can feel Larry roll over in his grave every time I click the "Start" button on my computer to stop it. Larry would never approve. Another early job at Radiation was the 540 Decom for NASA. Bernard was primarily responsible for the innovative design of this early stored program device that easily handled the complex formats of the PCM telemetry for the Gemini and Apollo programs.

For a number of years, Bernard and I went to the Sebring sports car race in his camper and had a great time admiring the latest cars and the latest trends in the infield spectators, particularly the females. Bernard always brought the food and cooked, and we ate well (and drank beer well, too). Bernard was never one to accept the conventional design in anything he bought or used. His first sailboat was the Indian River Sloop, which used bilgeboards rather than the conventional centerboard. His favorite auto was his Porsche 912, the four cylinder version of the much more popular six cylinder 911. He built his own truck-mounted camper, and took his family on an epic tour of the US in it. He designed and built the house on Whitmire Drive. Every feature was carefully planned.

Bernard took many trips, some probably too adventurous for his family at home.

The late Ralph Johnson, a neighbor and fellow Radiation employee, writes of a trip to the Bahamas:

Bernard and I went out the Fort Pierce inlet, heading for West End G.B.I. early one morning. My S22 was powered by an OMC IO. As we left the inlet, following a 32' Bertram, we took a wave over the bow that damaged the windshield. We found the sea condition such that we could not keep up with the other boat. We lost the ground plate for our ship to shore radio, so we couldn't communicate.

We spent all day fighting the wheel to hold our compass course. Late in the afternoon Bernard noticed a change in the color of the water. The depth finder confirmed we had reached the flat well north of our destination. With a calmer sea we headed south at full speed. Upon our arrival at West End we found George Shaw and the other Radiators very glad to see us, it was around 9 P.M. We were finally able to phone home to let our wives and families off the worry rug.

A long-time mutual friend was the late John McQueen. John McQueen was first a marketeer for Ampex, and later joined Radiation/Harris. He was a regular member of the lunch bunch. He wrote:

So many memories of wonderful times spent with Bernard, at his and Georgene's home, at the Melbourne Yacht Club, in Bimini, at Sebring, lunches with the clan at Radiation, Inc., on the sloop Eland, and just hanging out. He was the only real "best friend" I ever had. He and his family were a large part of my life.

Some of Bernard's oldest friends were Buck and Sara Jean Benner. Sara Jean writes:

Buck and I have a lot of fond memories of Bernard that we could relate, but it would fill a book. Bernard loved the Mills Brothers. My worst memory was at one of his parties where he mixed me a "Moscow Mule". At that time I did not drink. It was so good I made him give me another one, which he did not want to do. I spent the rest of the night locked in the bathroom passed out. Bernard was so sweet, he took me home, pointed me to my front door and hoped I made it to bed. I did, and had to go to work the next day. He even came to my work to check on me the next day.

Buck writes:
Bernard and I go back a long time ago --- 1939.

We were in the 4th grade, Alexander PS, Macon, Ga.. Our lives have been intertwined ever since--- all good memories. Probably more than anyone, I owe Bernard the most, not money, gratitude. I had just been discharged from the Army and was back in college at Georgia Tech. Bernard had a big party at his house--his mother was there, not as a chaperon but just having a good time. She was a great lady. Anyway, Bernard had a date with a good looking gal who caught my eye immediately. I got her phone number from Bernard and we were married about a year later. Still married to her.

Thanks again Bernard, you did good.

Walt Fredrickson Remembers

Growing up in Melbourne, and having an early interest in radio and electronics, I was generally aware that Radiation had been founded and occupied some old Navy buildings at the Melbourne airport.

Upon graduation from the University of Florida, and receiving an Air Force commission, I was faced with a few months before reporting for active duty. So, the logical thing to do was to start work for Radiation in June 1957, in Meryl Burns's section, along with Perry Knight, Ed Claggett, Stan DeShazo, et al. We were in Building 7, the old Naval base laundry, and watched the frogs and snakes crawl in through the scuppers at night. My project was a data recording system for rocket sled telemetry, whose vacuum tubes burned and shocked me many times. This was a great group of dedicated engineers, technicians and assemblers who taught me early on the value of hard work and project teams.

Fast forward to Spring 1960, when I was faced with staying in the Air Force, converting to government civil service or seeking an industrial position. My wife, who had a taste of Florida sand in her shoes, opted to return to Radiation, and we were sent to the Orlando-McCoy Instrumentation Division to build a PCM data display system for an aerospace company. Again, this was a great group of people, but the division was essentially competing with the Melbourne operation, and seemed likely to be consolidated with Melbourne in the near term.

Returning to Melbourne in 1961, I was assigned to an Airborne Division project with Don Sorchych to build the telemetry package for the Nimbus satellite. Don was an exceptional engineer, and clearly headed for management positions. Our mechanical packaging technology was pretty primitive, and I remember wincing each time we compressed the circuit boards into the aluminum box! Fortunately, it all worked, and turned out to be a very reliable subsystem.

Moving next to Ground Data Department, we built several systems for data processing and recording the Nike Zeus telemetry data, and then took on a major data processing system for North American Aviation's Saturn 5 rocket. Lots of fun, good stories and great people, including Sam Hersperger, Frank Lewis, John Lemasters, Howard Thrailkill. Installing the system in Downey, CA gave me a taste of smog-laden Los Angles, a good incentive to complete the installation and return to Florida. We subsequently used some of technology to build a line of modular, "standard product" multiplexers, coders and recording units.

After the 1967 purchase of Radiation, Harris started an effort to find commercial application's for Radiation's technology. Sy Noon, the Harris corporate VP of Engineering, assigned Paul McGarrell to Melbourne, and I was selected to help him develop a business plan for applying electronic technology to the graphics arts market – specifically the pre-press area of newspapers and magazine publishers. This began a serious 10 year effort to address this market with a complete line of pre-press equipment for the capture, editing and photo-composition of news and advertising material. Some corporate "seed money" and great mentors such as Dick Tullis, Dr. Boyd, Jack Hartley, et al made this effort a success through the decade on the 1970's and early 1980's, although also pointed out the great difficulty of commercialization of defense technology. I was very fortunate to participate in this venture, and we tried to capture some of the best practices and insert them into subsequent divisional programs throughout the corporation. All in all, a great experience with some truly talented and dedicated people!

Memories of Gantt Hamner by his family

One story Gantt always loved to tell was the "gator" story. Ever wonder why he always says "See ya later alligator" and "After a while crocodile"?

On the bank of a creek, Gantt unexpectedly hooked an alligator using his favorite lure. Not wanting to lose his prized lure, he straggled, lost his footing and slipped into the water and found the gator attached to his chest; nostrils flaring just underneath his chin. The smell of the gator's breath made him cringe. The gator's sharp claws dug into his sides until it burned him. The croc's sharp nails gave him scars for years. He managed to get to his feet but the gator was still firmly attached to him. Since he was alone, he was in a dangerous predicament.

Gantt had to think fast what to do, he decided the best thing to do was to shake his whole body and scream as loud as he could to startle it and to let go and run away.

Amazingly it worked, he'll never forget that day and those words he screamed. "See Ya Later Alligator", "After a While Crocodile"!

Dad specifically requested this saying be put on his urn - he used these phrases daily whenever a visitor or family member had to leave. He will be remembered for his many tall tales, his crab and shrimp parties, inventions and his good humor.

John T. Hartley Remembers

John T. (Jack) Hartley retired from the Harris Board of Directors in March 2002. Having reached the age of 72, the mandatory age for retirement from the Board, he concluded a career and an association with the company that spanned 46 years. He was CEO of the company from 1986 to 1995 and a member of the Board from 1976 until his retirement in 2002. His entire career was devoted to Harris and Radiation, Inc., the Melbourne-based company that merged with Harris-Intertype in 1967. He rose from the engineering ranks to become the CEO of a Fortune 500 company. Included below are some of the highlights of his career.

Jack was born in Jacksonville, Florida, in 1930. He and his sister were raised by his mother, who was divorced. The Great Depression had devastated the U.S. economy and there was no extra money in the single-parent Hartley home. "As a high school senior, I had little hope that I could afford a college education," he recalls.

Then one of life's "turning points" occurred. It was 1946 and the U.S. Navy was sponsoring academic competitions for four-year college scholarships. The Navy hoped to attract students to careers as naval officers, following the military downsizing post-WWII. Jack, who describes himself at that time as a "smart but unmotivated" student, won the competition and a four-year ride at Auburn University in Alabama, all expenses paid.

"I would have preferred to stay in Florida and attend the university at Gainesville, but the Navy selected the school," he says. As it turned out, Auburn was a good choice. "It was a smaller school with less affluent students, many from surrounding rural areas of Alabama. It was a comfortable fit." Jack graduated with a degree in Chemistry in 1951 and married his wife Martha, also an Auburn student, five days after graduating.

He was commissioned as an ensign in the Navy and served on a troop carrier during the Korean Conflict. At the end of his three-year Navy obligation, Jack struggled with the decision of whether to make a career of the Navy, where the pay and benefits were good. Another turning point. He decided to return to Auburn to further his education and pursue his passion for electronics. He earned a degree in electrical engineering, enrolled in graduate school, and supplemented his GI Bill with pay as a teaching associate.

Prior to graduation, Jack was contacted by several high-profile firms in the Northeast, including RCA, GE, and Westinghouse. It was 1956; American industry was beginning to ramp up and the Cold War was creating a demand for advanced electronics. Jack's educational background, coupled with his maturity as a military veteran, made him a valuable commodity in the emerging electronics industry.

"My hair was prematurely gray, so I looked older than I was," he says. "At the same time I was preparing for a round of interviews in the Northeast, I had a chance conversation with a stranger about a small start-up electronics firm in Melbourne, Florida – Radiation, Inc. I loved Florida and hoped we would end up there one day, so I made the trip to Melbourne." He never looked back.

The president of Radiation, Homer Denius, asked Jack to start the day he arrived for the interview. Although Jack did take the time to follow up with the big firms in the Northeast, his heart and mind were already at work in Melbourne.

When he joined the firm in 1956, Radiation had approximately 500 employees and sales of $2.2 million. The company had been started in 1950 and was providing digital telemetry equipment that gathered and transmitted on-board information from missiles being launched from nearby Cape Canaveral. The company had contracts with the Department of Defense and the National Security Agency. Jack was right at home; he had been working on digital electronics at Auburn.

Jack joined Radiation as a senior engineer and was soon leading the pursuit of some big programs, one in competition against the mammoth ITT. "We lost that contest, and I had the feeling that I had just gotten my first taste of the political side of government contract pursuits," he says. He was named VP of Advanced Programs in 1960. His job was to pursue new business and he began, as he describes it, "the transition from being an engineer to being a manager."

In 1964 Jack was chosen to run the company's fledgling microelectronics business, the predecessor of Harris Semiconductor. "Radiation was by far the smallest company—by a factor of 10—to enter into the integrated circuit business," he says. The technology was difficult and demanding. Initially the company wanted the circuits to support its digital telemetry products. On Jack's recommendation, the company expanded operations to become a merchant supplier, providing integrated circuits to other companies.

In 1967, Radiation, Inc., merged with Harris-Intertype, an international leader in printing equipment. The management of Cleveland-based Harris-Intertype envisioned a world in which printing and electronic technologies would converge. That vision never quite coalesced, but the marriage was successful. "Harris-Intertype was well managed and stable and, for the most part, they were not intrusive in the day-to-day operations in Melbourne," Jack says.

Only once in the years that followed did he come close to leaving the company for an attractive offer at a large electronics firm. That was the last turning point. He decided instead to lock his career to the fortunes of Harris- Intertype and the Melbourne operations that he had grown to know so well.

When the president of Radiation, Inc., Dr. Joseph Boyd, moved to Cleveland following the merger, Jack ran most of the Melbourne operations. He was promoted to VP-GM of the company's largest electronics division in 1968 and became a VP-group executive in 1971. The company's name was changed to Harris Corporation in 1974. Jack was elected executive vice president and a member of the Board in 1976 and president of the company in 1978. That same year, Harris Corporation moved its headquarters from Cleveland to Melbourne as the "intellectual focus of the company had shifted to Florida," according to Dr. Boyd. "It was clear that we were going to become an electronics company, and I wanted to be at the center of gravity."

Jack was elected CEO in 1986 and Chairman of the Board in 1987. During his tenure at the top of the company, sales rose almost 50 percent to $3.4 billion. The Semiconductor Division was more than doubled with the acquisition of GE Solid State; Lanier Worldwide was created with eventual sales of almost $1 billion; commercial communications businesses were added and sales expanded significantly in international markets; and the government business added to its base with important new contracts from non-DoD agencies.

Throughout his career, Jack has believed strongly in the importance that technological superiority has played in the success and prosperity of this country. "The U.S. has always been good at the intellectual part of the business equation. We may not always be able to compete in certain areas of manufacturing or mass production, but it is very important that we maintain our technological leadership." He notes that development within the government- industrial partnership has fueled most of the technological advances that the U.S. currently enjoys. "Technology is important to the defense of our country and the preservation of our way of life."

Jack has served as Chairman of the Board of Trustees of the Florida Institute of Technology in Melbourne, and he remains an avid golfer and has grown a prize collection of orchids.

Richard (Dick) Henry Remembers

I joined Radiation Orlando at the Pinecastle AFB where we had a company with access to the runway to outfit drone airplanes for testing the Bikini Atoll atomic clouds. Julian Scott, Scotty, was there then. After graduating at RCA Institutes in NYC I joined in Sept 14th 1959 and worked until I retired in 1999.

I live in Ocala now for 12 years and still do things like cut plastics with my industrial laser, mostly for quilting templates in my wife's work. She is Carol and is still working, teaching quilting, judging quilt shows, wrote a book, etc.

Remembering Art Herbert

From his obituary:

George Arthur "Art" Herbert, 83, longtime resident of Florida and well-respected venture capitalist, died on Saturday, February 20, 2010 from heart failure. He was born May 10, 1926 to Carl Morse and Jessie Freestone Herbert in Point Pleasant, NJ. He attended the U.S. Naval Academy in Annapolis, MD from 1944-1947. During the Korean War Art served on numerous ships, including the USS Montrose (APA 212) and the USS Doyle (DMS 34). During this time, he married Rebecca "Shelly" Sheldon Whitemarsh (d. 1999) with whom he had four children. He resigned from the Navy in 1954. In 1956, having graduated from Harvard Business School with an MBA, Art became a senior executive of Radiation Inc. in Melbourne, FL. In 1968 he turned his focus to helping provide growth capital to small companies, starting with co-founding Electro-Science, a venture capital firm. He later helped establish the National Venture Capital Association in Washington, DC. In the past two decades, he formed CEO Advisors, where he offered guidance to entrepreneurs developing unique high tech companies. He sat on the board of directors for Stryker Aluminum Yachts, Utah Medical Development Corp, Capital Financial Systems Inc., Vision Images, Inc. Cynris, Inc., ASTIK, Inc., Autonomous Technologies, TECHNE Corp and Blue Orb, Inc. Art had a life-long appreciation of the sea and was an active ocean sail racing crew member, including supporting the development and campaigning of Heritage, a 1970 Americas Cup Contender. He valued community service and was a member of Kiwanis International, Masonic Order, Scottish Rite, and Shriners International. He was also a member of the Florida Native Plant Society. A passionate wine enthusiast, Art focused on developing viniculture in Central Florida instead of subdivisions. He was a founding partner of Seavin, Inc., which owns Lakeridge Winery and Vineyards in Clermont, FL, and San Sebastian Winery in St Augustine, FL. Survivors include his wife, Marilyn (Mahaffey) Herbert of Edgewater, FL, whom he married in 2003; sons, George Arthur Herbert, Jr. of Lovettsville and Richmond, VA, and Ross Charles Herbert of Melbourne; daughter, Rebecca Angela Herbert Thompson of Charlotte, NC; stepdaughters, Kelley Swain of Jacksonville, FL; and Kristen Grimm of Washington, DC; nine grandchildren; and four great-grandchildren. A memorial service to celebrate Art's life will be held Saturday, March 6th at 2 p.m. at Saint Michael's Episcopal Church, 2499 N. Westmoreland Dr, Orlando, FL, 32804. To honor Art, the family asks that in lieu of flowers, donations be sent to the , , Pulmonary Fibrosis Foundation, Florida Native Plant Society, or the SETI Institute.

Jack Johnson Remembers

This story begins in 1961, fairly late by Radiation Old Timers' standards. But there are a few things still in store for those interested in pre-Harris history.

I was a young engineer, working for Collins Radio World Net Division in Dallas, when I first heard of Radiation, Inc. We bought one of Radiation's spiral scan feeds for an antenna we were building for ARPA. With its big, whirling, fiberglass cones, it was a design guaranteed to attract a mechanical engineer like me. My next encounter came a year later when Dick Baker showed up in Dallas, at what was euphemistically called a "staffing center". It was really a raiding party on Collins and Texas Instruments, and I was curious. Dick invited me to Melbourne for an interview, and I accepted. I figured that if Radiation had other characters as interesting as Dick Baker, I wanted to meet them.

My first adventure occurred before I even got to the plant. They had me fly into Orlando's McCoy field, and drive to Palm Bay via U.S. 192. The night was late and it was very dark. Everything went fine until somewhere between Holopaw and Deer Park, when I became unsure I was still on planet Earth. There were no signs of civilization anywhere. In those days, 192 was only two lanes, and the roadbed was only a couple of feet above the St. Johns river floodplain. The bulrushes grew right up to the roadway, and were tall enough that the stars could only be seen in a narrow corridor directly overhead. By the time I became aware of all this, it was too late to turn around and go home. There were no road shoulders, and literally no place to turn a car around. I know, because I kept looking all the way to Melbourne. My spirits continued to sink when I had to cross the old wooden bridge at Lake Hell'n Blazes. The desolation I had experienced made the drugstore at Minton's corner seem like the center of a major metropolis.

Things seemed a little better the next morning when I reported in at Radiation's new campus (three buildings) on Palm Bay Road. The headquarters building had goldfish in a pond in the lobby. Imagine that! Goldfish! I'll bet that was George Shaw's idea.

My interview was with Harvey Bush, the Section Head of the Mechanical Engineering Section in the RF department. Harvey told me that Pres Shrader was transferring out to the program management department, and he needed an engineer to take his place. He also introduced me to several of the ME's in the section, including Jim Perkins, Bill Corry, Vic Sakryd, Dave Balser, and Cliff Schoonmaker. Harvey offered me the job as a Senior Engineer. I didn't think much of Melbourne/Palm Bay or, for that matter, the state of Florida. But I did like the company and these people, so I accepted.

I reported to work at Radiation on February 5, 1962. Also arriving that same day was Ed Rabine, who had just been recruited from Convair in Arlington, Texas. We met before we got our badges, and went on to spend our entire careers with our new company. Of course, neither of us could see that then, and we expected to be around for only a year or two.

I got to my desk in the old Building 1 about 10 AM that morning, got my first work assignment before 11, and was hard at work before lunch. From that morning, the pace was breathtaking, and it never let up for the remainder of the decade.

That first assignment was the structures subtask on the TAA-2 program. We built two 85 foot diameter telemetry tracking antennas, one for each of the country's missile ranges. One was installed on Grand Bahama Island, the other at Point Mugu, Calif. Rohr Corporation provided the pedestals and dishes, and we built the feeds and feed supports. Bill Corry introduced the 8-spar feed support that we would use on every large antenna thereafter. Since we mechanical folks bought or subcontracted everything in those days, I got to know Al Sykes, Larry Lanham, John Diaz and Nelson Fox very well.

The next assignment was my first exposure to the signal intelligence world. The contract was simply called 6729, and it called for two 40-foot diameter tracking antennas. These dishes were mounted on the

Radiation Telscom pedestal, also designed by Bill Corry and Kurt Albrecht. The pedestal was a little small for the big dishes, so the antennas were housed in geodesic radomes to protect them from ice and wind. One system was installed in Caribou, Maine, and the other in Spokane, Washington. Much of the field work was done by Gordon Kirkland, Eddie Nelson and Joe Young.

The next job was a biggie – Stonehouse/Bayhouse. This was another sigint contract, and it was even more deeply classified than 6729. One end of Building 1 was sealed off for the project, and we had to change badges to go through the only door. (This door separated the project from the restrooms, so we all had to put nature's call on hold while we exchanged our inside badge for our outside one.)

This program was actually two separate systems installed on the same site in Asmara, Ethiopia (now Eritrea). Stonehouse consisted of an 85-foot diameter antenna on an "X-Y" pedestal. It was state-of-the art in all respects, and had interchangeable feeds and subreflectors to cover a very wide range of frequencies. In a similar arrangement to TAA-2, Rohr built the pedestal and reflector, and Radiation provided the feeds, feed support and subreflectors. Marv Livingston designed the feeds, and Chris Catsimanes designed the hydraulic drive servo system. I was responsible for the overall structures subtask.

Bayhouse consisted of a 150-foot dish on a wheel-and-track pedestal. The structure had been originally designed by Stanford University, for use as a radio telescope. It rolled around on a 110-foot diameter railroad track. Bill Corry made several improvements to the design, including an adaptive elevation drive mechanism. This system was much more simple than stonehouse, but it had a lot of gain. Dave Balser had the structures subtask.

We essentially finished the stateside part of the program by April, 1964, when we staged everything in San Diego, and loaded into a chartered transport ship bound for the port of Massawa. By "everything", I mean everything we could think of that we could possibly need for the following year. We included Conex containers loaded with throwaway household furnishings for the dozens of families and children that would accompany the Installation and Test crew. We even bought and shipped a Bluebird school bus, to transport the Rohr ironworkers to and from the work site.

While the ship was at sea, we began to move practically the entire program to Asmara. Pres Shrader had already been on site for several months, doing the site clearing and arranging for lodging and transportation for everyone. He even had to arrange to have several bridges reinforced on the road from Massawa to Asmara. I had been promoted to Lead Engineer by then, and Jack Schwartz and I and our wives were the first to arrive and join Pres. There were no housing units yet available, so we had to live the first nine weeks in the Ciao Hotel in downtown Asmara. Jack and I were working long hours every day, so Delores and Anna were left on their own. It must have been truly awful, but they didn't complain. They knew things would be better when the apartments were completed and the Volkswagens Pres had ordered were finally delivered.

Stonehouse/Bayhouse was a real challenge, technically and logistically, and the on-site work and the follow-on O&M contracts presented new and unique problems almost every day. The adventures of the Asmara team are far too numerous to describe in this summary. In addition to those named above, Some of the people who worked on site include Dick Wilson, Jack Davis, Al Fritz, Earl King, Art Lovejoy, Gerry Edwards, Joe Pepin, Jim Turner, Jim Crenshaw, Moe Oshier, Jim Hunnings, Eddie Bannister, Ed Rabine, Van Murray, Tony Barile, George Hennesee, Wayne Baker, Arlie Wheeler, Manny Vigliotti, and later, Doris Nobles. Also on this remote site we first met several people who would come to Radiation later. These include Paul Thompson, Del Wiese, Don Branum and George Doliana. This group and their families have forged bonds which have lasted for nearly a half-century, and those who still live have regular reunions to re-tell the stories of the project.

The foregoing narrative brings us only up to 1965. Upon returning from Asmara, I found that the world had begun to see me as a manager rather than an individual contributor. (I think I was the last to notice.) Harvey Bush put me in charge of the Structures Group within the ME section, preparatory to Structures becoming a separate section. The first two big contracts to come along that year were the AN/TSC-54 (Mark V) and the Altair program. Both programs were technically very difficult, and both had customers that were skeptical that we could do the job. Fortunately, we had two brilliant mechanical designers to apply to the task. John Mazur worked primarily on the Mark V, and Bill Corry led the design team on Altair. I bore the brunt largely on the political side (or so it seemed).

The Mark V was a very different undertaking for most of us. It was to be the first generation of transportable tracking systems for joint use by the military for the Defense Satellite Communications program. To meet all tactical requirements, it had to be towable over very rough terrain, flown in by military helicopter and dropped, and transported in a single C-130 airplane, along with a generator and 48 hours of fuel. These requirements meant that the antenna module had to weigh less than 6000 pounds, survive a three-foot drop onto concrete, fold up into a package less than 8 feet wide and high, and be fully deployed and operational in less than two hours. In addition, it had to be producible into hundreds of identical units – all things we had never done before. In the end, we did all of these things. Unfortunately, the Mark V was doomed by the frequency plan adopted by the government before we ever began. The frequency bands chosen for transmitting and receiving were just the right distance apart to produce a phenomenon called "third order intermodulation distortion. These intermods prevented the Mark V from carrying the multiple communication channels required. Only the first 13 were ever built.

The Altair system was a CW radar system, another first for Radiation. It was based on a 150-foot diameter antenna, and would be installed on Roi-Namur in the Marshall Islands. Its purpose was to track re-entering warheads at the end of the US Western Test Range. In this case, the prime contractor was Sylvania, and Radiation had the subcontract for the antenna, drives, feeds and RF transmission lines. The first big problem we had was winning and keeping the contract. The customer's technical advisors, the MIT Lincoln Labs, didn't want us. We had proposed a design-from-scratch wheel and track antenna, based on our experience with the Bayhouse program discussed above. Lincoln wanted the government to buy a scaled-down version of the 210-foot diameter antennas used by JPL on the NASA Deep Space Instrumentation Facilities. This design cost ten times as much as ours, so we prevailed. Lincoln continued to try to shoot us down throughout the life of the program. They made us prove everything. That cost a lot of money, but was probably good for the end product. Jack Davis was the Program Manager, Bill Corry designed the structure, and Til Sciambi designed the huge antenna feed . Pres Shrader was the site manager at Roi-Namur. Howard Sund supervised the structure assembly. I spent two months on site, doing the optical alignments on the track and the two-wheel trucks on the four corners of the pedestal. After Radiation completed its work, Dave Balser left the company and joined Sylvania for the Operation & Maintenance contract. Dave, Corky, and their two sons stayed on Roi Namur for nine years before returning to Radiation/Harris. The antenna was commissioned in 1968, and is still in operation today as a radio telescope. We were really tickled to prove Lincoln Labs wrong.

One other thing happened along the way. In 1966, I finally finished my graduate work at a little school call Brevard Engineering College. They presented me with a Masters Degree in Space Technology on a Friday. The following Monday, it was announced that the school was renamed the Florida Institute of Technology.

Also in 1966, I passed my professional engineers exam in Jacksonville, and became a licensed professional engineer in the state of Florida. That seal didn't do me a bit of good, but I felt like I had accomplished something worthwhile. It also improved the appearance of my resume in Radiation proposals.

While Mark V and Altair were approaching completion, something interesting was going on back in Palm Bay. Rumors of our merger with various companies were running up and down the halls, but nothing had yet happened. In the meantime, our management had decided it was time to move Program Management out of Engineering and make it a separate organization. This was the beginning of the matrix management system still in use today. Harlan Carothers was chosen to lead it, and a dozen or so people were selected as charter members of Program Management. I was one of them.

My first assignment in this newfangled department was to manage the in-country segment of the UK-60 program. What we called UK-60 was a satellite telemetry and command system for the United Kingdom Skynet satcom program. Skynet was a British version of our Defense sitcom program, and they bought it through the US Airforce Space Command. It was based on a 60-foot diameter scaled-down version of the Altair antenna (thence the term UK-60). Radiation's contract had an unique clause, which required us to procure 50% of all the material from sources within the United Kingdom. That part was my responsibility for the first part of the program. Larry Lanham and I, along with our families, departed for England in March, 1968. Our job was to develop a supplier base from scratch, finding credible sources for most of the materials and high-tech equipment we were going to need. That task required us to travel all over the country, and it was a lot of fun. Larry and I would meet every morning on the 6:05 train from Farnham, Surrey to London. From there, we would commute to wherever we were going that day.

Another part of my job was to get the Oakhanger, Hampshire, site ready for the installation and test phase, which would begin in the fall. The British Ministry of Public Building and Works, overseen by the RAF, provided the civil works. That task was also very interesting for me, and we all got a lot of laughs over differences in British and American technical terminology.

While Larry and I managed the in-country activities, John Cardinal, reporting to Harlan, managed the stateside portion of the program. Jack Acker was the assistant program manager. Don Shingler was the System Engineer, and Jim Perkins ran the structure design team. Andy Juettner developed the tracking system, and Charlie Vilardebo was responsible for the tracking software. Eventually, Jim, Andy, and Charlie moved to England to do the installation, along with Howard Sund, Bill Brophy, Roy Carter, and Al Beaupre, who ran the System Test function.

When the majority of the program had moved from Palm Bay to England, Harlan made me the overall program manager of UK-60. It was the first time anyone had to manage an R-110 from afar. Fortunately, I had Jack Acker as my anchor man at the plant. He made sure upper management was kept informed, and that I made all of the management decisions required of me. Shortly after this reassignment, Harlan left his position, and my new boss was John LeMasters. Although he and I had graduated from Georgia Tech in the same class, we had never met. (EE's didn't fraternize with ME's at Tech.)

By this time, Radiation and Harris-Intertype had merged, and we were now Radiation, a subsidiary of Harris-Intertype. I came home in September of 1969, got promoted to Engineering Section Head, and reported to Gene Whicker in the RF Department. I also moved into a changing culture, directed from Cleveland. But that's another set of remembrances.

Ralph Johnson Remembers

I interviewed at Radiation in the early fall of 1950, looking for some interesting work in Florida. I was hired, and reported for work in January of 1951, as employee # 19. At first I lived with several other Radiation engineers in a rented house off US 1 in Palm Bay. A few months later, I married Dot and we lived in a rented house in Indialantic.

Radiation was formally incorporated in August of 1950, at the law offices of Appleton Rosseter, with Homer putting up $20,000 and George $5000.

Bank of Melbourne and Trust Co (C. H. McNulty, president), the only significant bank in Melbourne, refused to do business with Radiation, not even to loan operating capital for US Government contracts. Radiation began working with Sun Bank in Orlando (Clay Dial). Years later Homer was instrumental in forming a competing bank in Melbourne.

Radiation sold about $100,000 worth of notes to finance operations, with relatives and Radiation employees (including myself and Charlie Keith) participating.

The first public sale of Radiation stock was in about 1955, underwritten by Johnson, Lane and Space of South Carolina, with some of the proceed used to pay back the note-holders. Later, a second issue was underwritten by Kuhn Loeb of New York.

Early jobs and programs

One of the rationales for locating Radiation in Melbourne was the proximity of the missile range at Cape Canaveral, and one of the early programs was a timing system for the Cape. This was an old vacuum tube system from White Sands Proving Ground, using lots of 6SN7's among other tubes. Our job was to deliver timing signals to remote camera sites. Various work was done on this system, including engineering, parts procurement, field service and repair.

One of the early jobs for Wright-Patterson AFB (Dayton, Ohio) was a FM telemetry transmitter, the Model 3115. Another early program for WPAFB was the design, test , and manufacture of the AN/UPM-19 test set for Identification Friend or Foe (IFF) systems. The UPM-19 operated at S-band, and included an oscillator, frequency meter, power meter, 6 audio tone generators and a vibrator type power supply. It had to operate over the wide temperature range of -55F to +100F. Manufacturing of the UPM-19 was done in a rental hanger at Herndon Airport in Orlando.

A shortage of work in Melbourne resulted in some engineers working for a time at Glen L Martin in Middle River, MD. They worked on the Shanicle guidance system (and test equipment), which was used on some versions of the Matador guided missile. Some Shanicle work was also done in Melbourne. [Shanicle was a LORAN-type system which used multiple rf transmitters to establish hyperbolic lines of position for the missile guidance system.]

Yet another program for Wright Field was for Rudder Pedal Dynamometers and Stick Force measurement instrumentation for installation in existing aircraft, with minimum modification., requiring clever mechanical design. Some old Strain gage amplifiers designed for this system (Model R-1040) were recently purchased on E bay by Nancye Meyers and are part of her Radiation memorabilia collection in Palm Bay.

Under another contract Wright Field sent us several experimental transistors from Bell Labs for evaluation. Efforts to measure their characteristics yielded inconsistent results. This was finally traced to varying humidity in Radiation's non-air conditioned lab, and imperfect seals on the transistors.

One project began routinely as a bid on a teletype machine which appeared in the CBD (Commerce Business Daily) listings. We won the job and it turned out to be for NSA (National Security Agency).

This was Radiation's first contract with the security community, a good customer in years to come. This was in the days when NSA was in an ex-girl's school on Nebraska Ave. in DC, before they moved to Fort Meade.

An unusual early product was a series of hermetically sealed circuit modules, filled with silicone oil for improved heat dissipation. One of these modules included a pump powered by magnetic coupling to an external motor.

Yet another system was an underwater sonar tow target for the Navy. This program was memorable because in testing the target broke loose and sank to the bottom of the Banana River off Dragon Point. (It was later recovered.)

Early Employees

Charlie Keith, Bob Bishop and Bob Dryden (all engineers from the University of Florida) were already at Radiation when I interviewed there in the fall of 1950. After I came to work, these were some of my fellow employees:

George Hedman, finance

Larry Culver, engineer. He had been Homer Denius' professor at Cincinnati, where Homer worked for Crosley during WW II.

George Anderson, engineer.

Don Gallentine, mechanical engineer.

Parker Painter, manager, ex Melpar.

Wally Bietel, technician.

James Ambrose Cramer, ex Raytheon and Melpar. He had a trailer full of WW II electronics he has purchased at surplus sales. Radiation designed around these parts and used them in breadboards.

Hans Scharla-Nielson, engineer and manger, ex GE. Created the Loop V and AN/UPM-19 antennas.

Jim Coapman, engineer. Worked on the ill-fated sonar target program. Later ran his own small electronic company (Missileonics, Inc.) in Melbourne. Later still, he worked for RCA at the Cape responsible engineering for the Range Instrumentation Ships.

Ed Dorsett, engineer.

Dean Ashwill, technician who worked on the UPM-19 and later moved over to the Orlando manufacturing operation when they started manufacturing the thing

A. W. Douglas. Worked in Financial planning, etc.

Ralph Bickford, first member of personnel department.

Bill Kercher, CPA, finance department.

Dave Howard, engineer and manager. Later died of a heart attack while playing tennis.

The first spin off from Radiation came in the late 1950's when Parker Painter, Bob Bishop, John Searcy, George Anderson and several others left to form Dynatronics.

Remembering George Lane

George E. Lane passed away recently in Tennessee after a long battle with the Alzheimer's disease. George was a counsel with Harris Corporation for many years and supported primarily the Government Communications Systems Division. He was instrumental in creating and leading the Harris Ethics/Business Conduct function. George's door was always open to anyone who needed advice. He played Santa Claus at the Children's Christmas Party for many years at GCSD – a role which came naturally to him. He was much loved and respected by all who knew him.

Bill Langford Remembers Compiled by Frank Perkins from a conversation on 5/20/2012

Bill first came to work at Radiation in August of 1960. At first he worked on the Minuteman program, and for a while he felt that 1290 (the Minuteman JA) was the only charge number at Radiation.

He then worked on the Nimbus PCM Telemetry Ground Station. [Nimbus was an early weather satellite, first launched in August of 1964. It was the second generation follow-on to Tiros, the first weather satellite to return useful data, in April 1960.] Bill did logic design on the Nimbus ground station, and also did some of the earliest software engineering at Radiation, as he taught himself to program the CDC 160-A computer in the System. [The CDC 160A was a 12 bit minicomputer built by Control Data Corporation.] Bill accompanied one of the Nimbus Stations to the site near Fairbanks, Alaska, and supported it during the flight of Nimbus 1. (Other Nimbus Ground Stations were delivered to GE in Valley Forge. PA, where satellite integration testing was performed.)

Next, Bill worked on the Lunar Excursion Module Checkout System. This was an automatic system for testing the Signal Conditioner Electronics Assembly in the Lunar Excursion Module spacecraft. It generated stimuli for the SCEA, and monitored the resultant SCEA outputs for accuracy. It included a DDP-116 computer. [A 16 bit, 3.4 microsecond cycle time minicomputer built by computer Control Co.] Bill presented a technical paper on the system at the Instrument Society of America conference held in Cocoa Beach in October of 1966.

Fellow employees that Bill particularly recalls are:

Leon Williamson
Russ Latch
Al Medler
John Lemasters   Ev Boyd
Dan Dupont
John Aruda

Remembering John Lemasters

From his obituary:
John N. Lemasters, III of Vero Beach, Florida, passed away suddenly on Tuesday, August 19, 2014, at the age of 81. John was born on August 3, 1933, in Akron, Ohio, to the late Helen (nee Crosby) and John N. Lemasters, II. He grew up in the Ocala, Florida area where he attended high school. He went on to graduate from Georgia Institute of Technology in 1958. During the Korean War, he served in the Army

as an instructor. John's professional career spanned over 40 years, in which he was President and CEO of Contel in Atlanta, GA, and Computer Products in Boca Raton, FL. Before that he was Vice President of Harris Corp. in Melbourne, FL, where he started the Satellite Communications Division. After retirement, he served on the board of several communication companies. He was an active member of Christ By The Sea United Methodist Church in Vero Beach where he played clarinet in the church orchestra. His hobbies included golf and piloting his airplane. In later years, he traded his airplane cockpit for his home flight simulator. Survivors include son Craig Lemasters (Dr. Laurel) of Atlanta, Georgia; daughter Lindi Lewis (Jan Lawrence) of Oakton, VA; grandchildren Will and Drew Lewis, and Steven, Kerry Anna, David and Corrie Lemasters; brother Steve Lemasters (Aloma); step-mother Shelby Lemasters; and four nieces.

Remembering John Lemasters

John was a section head at Radiation when I came to work. He was one of the original founders of Systems Engineering Labs in Fort Lauderdale.

From his obituary:

A resident of Vero Beach, he passed away suddenly on Tuesday, August 19, 2014, at the age of 81. John was born on August 3, 1933, in Akron, Ohio, to the late Helen (nee Crosby) and John N. Lemasters, II. He grew up in the Ocala, Florida area where he attended high school. He went on to graduate from Georgia Institute of Technology in 1958. During the Korean War, he served in the Army as an instructor. John's professional career spanned over 40 years, in which he was President and CEO of Contel in Atlanta, GA, and Computer Products in Boca Raton, FL. After retirement, he served on the board of several communication companies. He was an active member of Christ By The Sea United Methodist Church in Vero Beach where he played clarinet in the church orchestra. His hobbies included golf and piloting his airplane. In later years, he traded his airplane cockpit for his home flight simulator. Survivors include son Craig Lemasters (Dr. Laurel) of Atlanta, Georgia; daughter Lindi Lewis (Jan Lawrence) of Oakton, VA; grandchildren Will and Drew Lewis, and Steven, Kerry Anna, David and Corrie Lemasters; brother Steve Lemasters (Aloma); step-mother Shelby Lemasters; and four nieces.

Frank Lewis Remembers

Frank I. Lewis became a Radiation Incorporated engineer in 1960, just after his graduation from the University of Florida. He recently retired as senior vice president and assistant to the president and chief executive officer of Harris Corporation. He recalls his early days at Radiation:

My parents lived in Eau Gallie, Florida, while I was attending the University of Florida. All of their friends told them that Radiation would be a good place to work unless you were married because if you were you would never see your wife. And I was married.

I interviewed with a lot of companies but there was something at Radiation that was hard to describe — a very high-spirit ed enthusiasm, an excitement about the place. Just how Homer Denius and George Shaw created that, I'm not sure, but it was there when I arrived in 1960.

Homer was basically marketing and George was engineering and that's the combination you need. So I decided on Radiation. The hours were long, but my wife and I worked it out.

If you were willing to work and had the ability, there just didn't seem to be anything to hold you back. Within two years I was writing proposals and committing the company to million-dollar offerings. After three years I was managing one of the largest ground equipment development contracts Radiation ever had.

It was not at all unusual on a Saturday morning to see George Shaw in his flowered shirt and sneakers, kibitzing with the engineers. No one said you had to be there on Saturday. You just wanted to be part of it. There was a solid sense of superior engineering capability that permeated the place. It was damn near self-disciplining. When a new engineer was hired, very quickly his peers got to know whether he was a good engineer or not If he wasn't a solid engineer, he was practically driven out of the system by his peers, kind of cast off to the side.

Our organization was project oriented, and you lived and breathed by the quality of engineering people on your project. Every project engineer wanted the very best engineers. It was like kids in the schoolyard picking softball players during recess. The best ones got picked first and they played. The others either sat on the bench or didn't get picked at all and left the company.

At a critical juncture early in Radiation's history was when we got in over our head on the Minuteman telemetry project, and Boeing came in and helped us get back on track. It not only gave us the discipline we needed; it gave us a relationship with Boeing that continues to this day. At one point, we probably had as many Boeing engineers in our plant as our own engineers. We had a superior technology in PCM telemetry that was critical, and they helped us make it a success. We were basically a technology company, and we acquired needed program management and engineering management skills.

When Joe Boyd joined the company in 1962, he perpetuated the excellence-in-engineering attitude. He was less flamboyant than Shaw. He was there on weekends mixing with the engineers, but not in a flowered shirt. He was very quickly accepted as a competent engineer and manager.

Here's an example of the drive for engineering excellence. One engineer was building a piece of airborne equipment on a test bench, but it wasn't working quite right. His department director was known as a wild man about reliability and quality, and was raising Cain with him about reliability. "This equipment has to work in any environment," he said. The engineer had a bottle of Coca Cola nearby, and his boss picked up the bottle and poured Coke all over the equipment. "Do you see?" he said. "I want it to work even if it's got Coke all over it."

The surviving electronic product lines today are essentially the Radiation product lines and the surviving management of Harris is basically the Radiation management, but the surviving name is Harris. We were, and still are, one of the leading suppliers of big satellite dishes for NASA, the military, and other government agencies.

We have probably delivered more satellite communication systems than any other contractor. We have always been on the leading edge of that technology. We have always had a sense that we are doing something important for our country.

Remembering John McQueen by Frank Perkins

Sadly, I recently received word that John McQeen had died.

When I met him, sometime in the 1960's, John was marketing for Ampex, the major producer of magnetic tape recorders. Radiation was a big customer for Ampex, the so we saw a lot of John. Eventually he came to work at Radiation, and he became a regular member of the engineering lunch gang.

John was interested in automobiles, among many other things. He and Bernard France attended the Sebring 12 hour race together many times. There was a story of them headed to Sebring early one frosty morning in John's vehicle, a big Lincoln. The heater controls were operated from engine vacuum and were not functioning. They disconnected the vacuum line and manually sucked on it to get the heater to work. Later, John had a Ford Cortina, chosen because it was fun the drive and had enough room inside to transport customers to lunch.

Thinking of John inspired me to meditate on and research the early Ampex tape recorders. Actually, my first contact with Ampex recorders was in my Army days. The Corporal missile, on which I worked, used a large Ampex recorder (it must have been a Model 302), to record analog telemetry for later evaluation. At Radiation, we used several types of Ampex recorders in systems we built. In the Minuteman ground systems, the FR 600 instrumentation recorder was used to record the serial PCM stream from the missile. It occupied a full 19" wide electronic rack. Another Ampex was the FR 400, one of the first recorders designed for computer input/output. One challenge was that computers wanted to read the data a block at a time. The recorder had to be able to start and stop within the inter-record gaps, and operated at high speed within the block. The FR 400 had mechanical arms to buffer the section of tape passing the heads from the large mass of the 10.5" reels holding the bulk of the tape. The FR 400 held up to 3600 feet of tape. At the typical width and recording density, this held around 10 megabytes of data. I recently purchased a magnetic tape recorder (albeit a hard disk) that holds a terabyte of data (ie, 1000 Gigabytes). My recorder is about 3" by 4" by 0.5", and cost under $100. The FR 400 was about 19" by 24" by 14" and certainly cost over $10,000. I have a PDF file of a brochure on the FR 400 that would require about half a reel of tape to store on the FR 400!

Dan McRae By Frank Perkins

Since he was so important in my life, I wanted to say a few words in memory of Dan.

Football Pool
He is best known, of course, for developing a football pool carefully designed to keep the maximum number of entrants in the running for as long as possible during the bowl season, but I want mention a few of his other accomplishments.

My Association
Many of you are aware that Dan and I worked on a number of projects together at Radiation and Harris. In these, Dan was the conceptual inspiration and I was the implementer. Dan was much more than a brilliant technical innovator: he understood the business and marketing aspects of making his inventions useful and feasible to the both the company, and our customers. His ideas had to satisfy the users needs and to be aligned with the plans, resources and inclinations of the company.

Dan had the skill to present an idea to people of a wide range of interests and sophistication. He could speak equations to mathematicians, economics to managers, customer relations to marketeers and applications to customers. He was skilled at instantly assessing his audience: he seldom talked down to you, nor over your head. He could even explain to me what to build.

TVSS
My longest running association with Dan(stretching from the mid 70's to the late 80's) was in the general field of modems. It all started though with a voice encryption scheme of Dan's. At an early presentation the customer told him that it would not work on a "real" channel: they had tried it and failed. They were so insulting in their assessment that Dan took it as a challenge and developed an adaptive equalization technique that made it work, and ultimately resulted in TVSS (Telephone Voice security System) which we sold to another (competing) customer. TVSS was notable to me because I took a prototype on a globe-girdling test trip to Europe, India and Japan.

16kb/s Modem

We realize that the the equalizer could also be applied to digital transmission on phone lines. At that time the state of the art was 2400 bps on common, unconditioned lines; complex modems were marginally capable of 9600 bps on expensive conditions lines. The military needed 16000 bps on dial-up lines for secure voice applications. This had been unsuccessfully attempted by telephone modem experts and our claims were ridiculed until we built a breadboard modem and demonstrated it. Even then our market was blocked by competing approaches until we incorporated the modem in a secure voice terminal (the Vinson Autovon Terminal, or VAT), and eventually sold over 1000, an unheard of quantity for Harris at the time. Field testing and demonstrations of the 16kb/s technology generated a lot of interesting trips. Dan, I and others stumbled through a series of trips through customs to a variety of military installations in Italy, Germany, England and Hawaii. Later there were Host Nation Approval tests in many other NATO counties. Potential export sales even took me to Peru, and a side trip to Machu Pichu.

HF Modem

Many of our demonstrations generated questions of the application of the technology to the HF radio channel. It turned out that the equalizer could not cope with the fading and multipath of this channel, but Dan developed a system that would, at least up to 2400 bps. Comparative link tests showed it worked on ten times the number of links than the current state of the art modems.

Again, testing trips were interesting: I flew to Ascension Island in the far South Atlantic in a droning Air Force plane; sailed with the modem on a Navy ship to Pearl Harbor, while bunking with engine room hands; and demonstrated in Korea while touring on a false passport (it's a long story).

We never sold many HF modem despite the technical successes, largely because of undercutting by the traditional HF division of Harris, but we sure had fun!

Thanks, Dan.

Remembering Roland Moseley
By A.B. Amis

Roland Moseley and I were friends for more than fifty years. We were across the street neighbors for maybe twenty of those years, sharing ownership in a lawn edger and things like that. Our two pet basset hounds were run over by the same school bus while chasing a cat, and I think I had to bury his for him – he was too tender hearted to do it himself. We partied together and worked together during the 1960s in what to me were the golden years of my career. I knew Roland as an outstanding engineer and proposal writer during my work association with him in RF, but he evidently also proved to be a good salesman later on in pursuing business from NATO and the FAA and others.

But there was another, more human, side to Roland that needs remembering too. Well, for one thing, Norma Jean Goode would tell you that he always found some excuse not to take her out to lunch on National Secretaries Day, but I was usually happy to stand in for him on that account.

But outside of work, Roland was always ready for a party. I recall one fine October day in 1964 Roland, Gene Whicker, Mel Cox, Guy Pelchat, Ray Penn, Al Jacobs, and I set out in two boats from Hatt's in Melbourne to find some clams for a chowder. We finally anchored in the river some mile or so north of Sebastian Inlet and stripped to bathing trunks and tennis shoes, grabbed a can of beer and started probing the muddy bottom of the river, feeling for clams. Guy had taken off his glasses so they wouldn't get lost, and kept us entertained by shuffling around with his eyes closed and singing dirty French songs while feeling for the clams. When we felt a clam, the procedure was to hold your beer up high in one hand while reaching under the water with the other hand, and then drop the clam into your bathing trunks so you didn't have to make so many trips back to the boat.

It seems like every trip back to the boat called for another beer, and since we weren't very environmentally conscious back in those days, we'd maybe throw our empty can back in the direction of the boat, and if it missed, no big deal. Before too long you could see a wake of empty aluminum cans drifting to the south, bobbing in the water and reflecting the sunlight – quite a beautiful spectacle, actually.

When we finally concluded we had enough clams, or else ran out of beer, whichever came first, we all got back in the boats except for Roland, who insisted on showing us his imitation of a whale – which consisted of floating on his back and spewing water out of his mouth. We finally got him back in the boat by threatening to go off and leave him, but he was still feeling "frisky" when we got back to Hatts landing and insisted on showing us his famous "Moseley barefoot ballet", skating down the slippery boat launch ramp on his bare feet, and oblivious to the fact that the concrete bottom was also covered with small razor-sharp barnacles. So by the time we got the boats out of the water and all got cleaned up for the chowder making party over at the Whicker's house, Roland's feet were so cut up that he wasn't able to walk. No problem, as Gene had somehow acquired an old hospital gurney like they use for moving people in and out of ambulances, so Roland got up on that and we'd push him around from place to place. But before long just pushing him around wasn't enough, so it became a game of dragging the gurney up and down Magnolia Avenue fast and making quick turns to see if we could throw him off. Poor Roland!

The chowder turned out great, and we all had such fond memories of that outing that 18 years later, in October 1982, we determined to have a renewal of "The Great Indian River Clam Digging and Beer Drinking Expedition" down at a place Guy owned on the Indian River just a few miles from the scene of our 1964 expedition. Mel Cox was charged with the job of coming up with a suitable monument to mark that occasion, and I don't know what charge number he might have used but he somehow got the Harris machine shop to engrave a brass plate commemorating the event and he attached that to a concrete base. I have a photograph of it right here. We all had an overnight party/campout down there at Guy's place, and the next day had a ceremony putting the monument in place. I wonder what ever became of it – the last time I saw it, it was beginning to be covered with weeds and vines.

We had several subsequent "Survivor" parties down there and at our place in Grant, but now there are only three of us Survivors left – Guy, Ray, and myself – and to look at the three of us, we're not doing all that good either!

Now I know you might have gotten the idea from what I've said that Roland was "bad to drink", and I guess he may have been back when we were all a little bit younger and wilder, but the truth is that it just didn't take very much to get Roland tipsy. Ray and Guy will probably recall one incident in a bar up near Ft. Monmouth -- I think we were chasing Mark V at the time -- where Al Jacobs was trying to drink Roland under the table, and even after secretly asking the bartender to switch him to just water, he still got so drunk on mostly water that Guy wasn't even able to pronounce his name – every time Guy tried it would come out "Roman Roseley".

My pet name for Roland was "Reggie". Some of you may recall the Reginald Van Gleason, III character on Jackie Gleason's TV show, an overweight playboy with a perennial cocktail in his hand. Well, my brother-in-law could never remember Roland's name and usually came up with Reggie instead, and that just seemed to be a good name fit so I adopted it and Roland answered to it.

I recall another incident where a few friends had gathered at our home late one afternoon for some partying, and Roland happened to be walking past and spotted all the cars and came in and joined us. After a drink or two I called Marita, Roland's, wife, and told her that I was Chief Mardell of the Indialantic police, and that one of my officers had picked up a man who claimed to be her husband walking along

Tampa Avenue and charged him with WWI – that's Walking While Intoxicated. She replied that this was probably her husband all right, but he really hadn't had all that much to drink. I finally let her off the hook and she came over and joined the party.

But in his later years, after some serious heart issues and multiple bypass surgery, Roland had really cleaned up his act – giving up both smoking and drinking, losing weight, and taking up an exercise program that included biking on the bike path all the way down to the Inlet sometimes. I really admired and respected him for that. When we last had lunch together back about six months ago he was just regaining strength from an extended hospital stay from those same heart related issues that had never completely left him, and he was looking forward to getting back to workouts in the gym, so his death really shocked and saddened me.

Rest in peace, Reggie.

Tribute to Harold O'Kelley 12/11/00
by AB. Amis

I loved this man!

The happiest times of my working life were the years spent working directly for Harold in the early 1960s.

It's hard to realize that was almost 40 years ago. Those were very special years for me, and doubtless for a number of other people that I know are here today. And they were seemingly special years for Harold too, because this is where he chose to return and spend the final months of his life, among old friends.

I'll always remember and treasure the lunches we've had together during the past few months, when Harold had Ray Penn round up a few of the old gang that used to gather in Harold's office for bull sessions every afternoon before going home from work. While we all knew that we were saying goodbye, the conversation at the lunches was always upbeat and filled with happy reminiscences about the good old days. We were able to pick up right where we'd left off forty years earlier, hardly missing a beat.

Harold's daughter, Betts, and I were talking the other day about what it was that had made those times in the 60s so special. She asked if it could have been, in part, because we were so good at the technology we were practicing, but I said no -- I don't think that was it -- I think we were good enough, but I don't think that was what made it special. Betts then ventured that maybe it was because people always knew where they stood with Harold. I believe she was getting much closer to the reason when she suggested that, and I told her I'd need to think about that some more.

The best I've been able to come up with is this: As Harold first began flexing his managerial wings, his low key and straightforward management style struck a particular resonance with an irreverent group of guys in RF, which enabled him to mold them into a team where everyone understood and was comfortable with his individual role. No internal political games were being played -- and if there were problems of any sort, then they got worked out informally at those after hours bull sessions -- where just as likely Mel Cox would insist on demonstrating that he could touch the tip of his nose, or his chin, with his tongue, or Guy Pelchat would show off his physical prowess by doing handstands on the back of a chair or offering to arm wrestle anyone present. I can't remember now what Harold's show-off stunt was, but he wouldn't have gotten away with not having some stunt to show! Being the boss didn't carry any special privileges, and that gang worked hard at keeping Harold humble, then and later.

I recall one particular incident where the company photographer had just made passport style photos of everyone in the company – probably for a new company brochure or something of the sort. Well, Harold's photo had turned out particularly awful – even to the point of being funny to some of us, so one of us, I don't remember who, had the photographer secretly make an 8x10 enlargement. Then one day while Harold was out of town we made a dozen or so Xerox copies of that enlargement and dispatched one of the planners down to the office supply store to buy a half dozen plain black 8x10 picture frames – intending to surprise Harold by having this awful photo peering down from the office walls of his staff when he returned.

Well, mischief like that hardly ever turns out just the way you had originally thought it would, so when we all gathered in a conference room somewhere to begin placing the photos in the picture frames, someone started playing around with one of the copies since we had far more than we needed anyhow. I think the first thing someone did was to cut out Harold's eyeballs and turn them upside down – creating the effect that he was ogling a secretary in a low-cut dress. Everybody immediately got into the spirit of the thing and started cutting and pasting and doing assorted violence to the photos to produce hilarious effects. One I remember lengthened his forehead, left him just a fringe of hair above his ears, and added a clerical collar. We labeled that one "Father O'Kelley". Harold had a reputation among us of being tight with funds, so we doctored up one photo to give him a foreshortened forehead, foreshortened upper lip, and no neck at all – and then painted dollar signs in place of eyeballs. On others we did more subtle things like just slightly crossing his eyes. We probably wasted a whole morning at this, and when we were finally laughed out and satisfied with our handiwork we hung the pictures and went back to work.

When Harold started on his normal rounds the next morning, checking in and stopping to chat briefly with various of his staff, he pretended not to notice when he encountered the first of the photos, but the second one was harder to ignore, and when he started finding them in every office and started looking at them more closely, he realized he'd been had! I don't think those pictures graced our walls for more than that one day, because we figured we'd better get them down before something bad happened to them. (Or to us.) I ended up taking the whole bunch of them home for safekeeping, and then some years later I believe Sarah talked me out of them.

Harold O'Kelly Remembers
(Excerpt from his book)

The year was 1957, over forty years ago. As we drove on U.S. #1 and approached Melbourne from the north the sights depressed me. Melbourne made me think of an imitation, run-down shabby Florida tourist town. On the left the highway ran next to the rotten-smelling Indian River. On the right ochre colored stucco gas stations covered with the dirt of many years decorated the two-lane broken asphalt highway. Interspersed were miscellaneous shacks offering the goodies of a Florida vacation: beer, hotdogs, T-shirts three for two dollars and inferior souvenirs made of seashell. The sparse vegetation amplified the heat of the sandy sunshine day. We passed the one-story frame structure, painted pale green, that inadequately served as the local hospital. In town center we encountered a run-down movie theater. We left-turned off U.S. #1 and crossed the Melbourne causeway centered with a swing bridge and lined by Australian pines. Thus we arrived at the Town of Indialantic where a dingy pink duplex on 3rd Avenue awaited us. From this dump, I reported to Radiation Incorporated.

Homer Denius and George Shaw—both engineers previously employed at Melpar—founded the company in 1950. They located their start-up at the Melbourne Airport because World War II buildings were available at reasonable rents and because it was in Florida. The beginning technology was Digital telemetry for aircraft, missiles and spacecraft. Later they expanded into transmitting and receiving equipment. The Federal Government was their target customer.

The Director of the RF Systems Division, John Downs, offered me a job as Project Engineer—the title described a pay grade rather than responsibility. My first assignment was on Project 1130, "Doppler

Navigation Radar for Army Helicopters." I acted as a systems analyst evaluating the spectrum of the signal back-scattered from the ground and returned to helicopter radar. This was a clue to me that my bosses, because of my prior professorship, looked upon me as a scientist rather than engineer. Even though the prototype equipment was tested in a fixed wing aircraft, it never was placed in production—the prototype did not work very well. Actually, the project ended ignominiously when the test aircraft made a bad landing, severely damaging the radar and slightly injuring Warren Wiener, our Project Engineer leading the effort. Next Downs assigned me to a project for the U.S. Weather Bureau.

Doppler radar displays are seen when we watch our local weathercast. The term "Doppler" (a man's name) means frequency shift caused by moving objects. County Mounties use Doppler when they catch you speeding in your automobile. When I joined Radiation, Doppler radar was not widely used. Then the military used it for measuring aircraft ground speed. The Weather Bureau theorized that Doppler radar might detect tornadoes. A tornado touching the ground picks up trash and twists debris with speeds guessed to be 200-300 miles per hour. Doppler Radar only detects a frequency shift from objects having a velocity component directly toward or away from the radar. Reflections from objects moving perpendicular to the radar-beam do not have a frequency shift. Therefore, a funnel cloud filled with swirling debris would show velocities from zero to, say, 300 mph. The Weather Bureau wanted to test their theory. Further, they hoped to record the velocity spectrum of a funnel cloud to determine actual speeds. I was the Project Engineer at Radiation designated to design and build the Weather Bureau Doppler radar. We took an old radar trailer, supplied by the government, and used it to house the electronics. We put two new dishes on the antenna mount, and made it into a X-Band CW Doppler Radar. (X-Band means that it operated at about 10,000 megacycles—megahertz—and CW meant that it sent out a continuous wave instead of pulses.) The acceptance test was to show that a BB shot directly at the radar from a distance of 100 yards was detectable. My radar passed easily. From this project I earned the nickname "Overrun O'Kelley." Marketer Myron Roebuck, who later became Vice President of Marketing, called me that. The fixed price was $30,000, and cost was estimated at $27,000, I brought in the finished product for $28,000. I still respond to the name when Buck calls me.

Dr. Joseph A. Boyd joined the company as Executive Vice President. One of his changes was to strengthen the pay grade titles of Radiation's technical personnel. An engineer could have a pay grade title and a functional title. In March 1959, I was placed in the pay-grade titled Associate Principal Engineer. At about the same time I was promoted to Section Head which was my functional title. A RF Division organization chart dated June 1959 shows me as Head of the Radar Systems Section and John Downs was Division Director. Managers under me were Jack Robbins, Servo- mechanisms and Electronic System Unit, Field Operations Unit under Vaudie Vice and Radar Systems Unit headed by Warren Wiener. As described before, Wiener's unit was working on navigation radar for army helicopters. The primary business of our section was to design and build large tracking antennas. Vaudie Vice installed them.

Once Daddy asked me an engineering question. He had never asked technical questions before, and didn't ask any later. He said, "Do television stations broadcast all picture sizes to accommodate the different sized pictures on home sets?" It never occurred to me that that could be a question on someone's mind. I told him that television receivers interpreted the signal into different picture sizes.

When I joined Radiation, Incorporated in 1957, I was in government electronics. Harris acquired the company ten years later (1967), and for a few years after that my job was still government business. Many of the programs I was involved in are well known and others not so well.

TLM-18 Tracking Antennas – Under my engineering management, we built sixty-foot diameter parabolic tracking antennas. These were installed on the Air Force Test range that began at Cape Canaveral Missile Launch Site. Radiation designed the TLM-18 and built and installed four. Initially it was thought that five units would be needed to cover the complete test range, but after three were installed it became apparent that they provided all the coverage needed. The three were installed at Cape Canaveral,

Ascension Island, and Antigua. Later we built and installed a TLM-18 at South Point, Hilo, Hawaii, to recover telemetry from missiles launched from Pt. Mugu, California. The prime function of the tracking antennas was to acquire telemetry data from missiles in flight, data such as status of various flight functions like engine cutoff, gimbal position of the in-flight steering system, heart rate on manned missions, etc. After the antenna acquired the flight telemetry transmitter, it would automatically follow the missile until it went below the horizon. The mechanism that provided the auto track capability was a rotating lens located at the parabola focal point. When the missile telemetry transmitter was off the neutral point (relative to the rotating lens) the amplitude of the received signal varied in synchronism with lens rotation. When this signal amplitude was other than zero, the antenna was not pointed directly at the missile. This error signal was used by the system to correct antenna pointing. This was a very successful program and was the beginning of our reputation as world class designers of telemetry tracking systems. Subsequent improved models of the TLM-18 were the Model R-1162, one each located at Vandenburg Air Force Base, California, and Kaena Point, Oahu, Hawaii, and one model R-6275 installed at Fort Monmouth, N. J.

TIROS – Another government program was a major prototype for weather forecasting, TIROS. Since 1960, television cameras have been used extensively on orbiting weather satellites. Vidicon cameras trained on the earth record pictures of cloud cover and weather patterns during the day. Infrared cameras record nighttime pictures. The ten Television Infra-red Observation Satellites (TIROS) launched by the National Aeronautics and Space Administration (NASA) paved the way for the operational satellites of the Environmental Science Services Administration, which in 1970 became a part of the National Oceanic and Atmospheric Administration The pictures returned from these satellites aid not only weather prediction but also aid understanding of global weather systems. Radiation built a sixty-foot tracking antenna to communicate with TIROS Satellites. It up-linked commands to the onboard cameras and received the down linked weather pictures. The infrared pictures were much more useful than the daytime regular pictures. The TIROS satellites worked at a low altitude (200 miles or so) so that any one picture showed a limited amount of earth area. By way of comparison, today's weather satellites are located at synchronous altitude and one picture can show weather patterns over almost one half the earth's globe.

MANNED SPACE FLIGHT - Astronauts were needed when the National Aeronautics and Space Administration (NASA) undertook manned space-flight programs in 1958, including those of Projects Mercury (1961-63), Gemini (1965-66), Apollo (1968-72), Skylab (1973-74), Apollo/Soyuz (1975), and Space Transportation System, better known as the space shuttle (1981). The early projects, Mercury and Gemini, used analog telemetry Since Radiation was expert in digital telemetry we did not have much to do with these early programs. We did, however, build some ground tracking antennas that were used in the Gemini project. The first man to orbit earth was Russian cosmonaut Yuri A. Gagarin, in 1961, and the first American in space was Alan B. Shepard, Jr., also in 1961, in a non-orbiting or ballistic flight. The following year, John H. Glenn (now Democratic Senator from Ohio) became the first U.S. Astronaut to circle the earth, and in 1963, Valentina V. Tereshkova of the USSR became the first woman in space. Three U.S. Astronauts (Neil A. Armstrong, Edwin E. Aldrin, Jr., and Michael Collins) manned the historic flight that landed Armstrong and Aldrin on the moon in July 1969. In 1983, astronaut Sally K. Ride became the first U.S. Woman to enter space.

The Apollo Command Module and the Lunar Excursion Module (LEM) were separate spacecraft but they worked together. The Command Module was designed to carry the astronauts into orbit and fire booster engines while in earth orbit in order to fly to the moon. The LEM was a tag along. Once in orbit around the moon, the LEM would separate and fly down to the surface. When the astronauts completed their mission on the lunar surface, they climbed back into the LEM and lifted off the moon's surface. The LEM rendezvoused with the CM. The astronauts then dispatched the LEM to crash into the moon. The LEM was designed to carry two astronauts. It had no capability to withstand the forces of reentry into earth's atmosphere. That requirement was met by the Command Module. We at Radiation won the contract to design and build the Pulse Code Modulated (digital) telemetry. NASA had special requirements on all the equipment designed for the CM and for the LEM. Since the mission was a manned space flight, reliability of the equipment was especially important. At Radiation, we met the reliability requirement by using quad-redundancy, which is a design that includes parts in such a way that if one part fails, there are three others available to take over. Thus each electrical part was backed up by three others, for a total of four thus, quad-redundancy. Radiation designers had many discussions regarding the question, "At launch how do we know that something has not failed." The thought was that if something had failed, the design caused a redundant component to take over. In quad-redundancy designed systems, a failure is not detectable until all four components fail. This is not a very likely situation. So the question is answered: "At launch, one cannot be sure that there has not been some failure(s)." The telemetry equipment for Apollo's Command Module and for the Lunar Excursion Module were quite successful. Our equipment flew on all the Apollo flights and some is still on the moon.

Shannon was born in the Melbourne Hospital January 15, 1966. The hospital was a small inadequate frame structure located on U.S. #1. Because of over-crowding Sarah's bed was in the hallway. At Hal's birth, February 1, 1967, the hospital board had built a new building away from the Indian River.

Early in my employment at Radiation I concluded that the best way to get ahead was to be involved with winning proposals for new business. A major reason for choosing Radiation over other companies was the fast growth of the company. A growing company created more opportunities for advancement. The faster the growth, the greater the opportunities for advancement.

Preparing proposals required that a document be written describing the method and designs we would use to achieve the requirements defined by a customer's specification. This required that we innovate. We created new designs and methods and described them. We made an estimate of what the work would cost. Based on the estimated cost, the company established a price for submission. Frequently, if the proposal was in the running, customers asked us to make a presentation on it. Usually, during a presentation our customer wanted to evaluate the persons who would do the work. I found the process of preparing a proposal exciting and fun. The innovative process was especially stimulating. Customer presentations were easy for me because of my experience in classrooms. Proposals to government agencies always had a specific due date. There was tendency to work up until the last minute. For me the frequent result was that I worked all night. At the time, I smoked. One disadvantage of working round-the-clock was that I smoked all the time that I was awake. I became proficient in proposal preparation. This stood me good in my quest for advancement. My experience in dealing with people was helpful in my career in industry.

In March 1961, I was promoted to Director of :he RF Systems Department. I had five sections under me: Antenna Systems, Microwave Design, Receivers, Electronic Countermeasures, and Mechanical Engineering. The Antenna Systems section was where I had worked before. The section was responsible for systems design, systems integrity, servomechanism design, and systems installation. . Microwave Design and Receivers Section had design engineers specializing in radio reception. Our mechanical engineers provided structural design for antennas. Electronic Countermeasures was a specialized unit not involved in the main business of the department. I tended to manage my department with a hands-on approach.

Mischa Kantor was an unusual man. Misch was a foreign born engineer who had worked at many companies. One day he showed up at Radiation asking to see me. At the time I was Director of the Surface Systems Division. I had never met the man nor had I heard of him. When I met with Kantor I found a man with an accent and somewhat older than me. He told me that he lived on his boat at the Melbourne marina. After checking around, he heard that I was the best engineering manager around and he wanted to go to work for me. After an extended interview, I told Kantor I would put him to work but he would be on probation for two weeks. If he had not proven his worth by that time he would have to go. Radiation management had closed our Orlando Engineering operation and transferred everything to Melbourne. Joe Boyd put all of the work under me. This included a particularly troubled project that carried the innocuous title "High Speed Printer." Our customer was Lawrence Radiation Laboratories, a division of the Atomic Energy Commission. The printer was intended to rapidly print computer output. The printer was capable of printing at 1,200 lines per second. Print speed was such that the paper, after folding, came off the print head and was automatically placed on a conveyor belt. To achieve this speed the printer used a paper coated with an electrical conducting material. The paper was moved under a row of tungsten styli wires. When an electrical pulse was applied to a stylus a black dot was made on the paper. The paper was 12 inches wide so there was a row of styli that same width. Spacing between the styli was a few millimeters. Therein lay the difficulty. With the paper moving rapidly past the styli, and electrical power being zapped to the styli, considerable heat built up. Also because the paper coating was burned in the process, a good deal of carbon was generated. As a result, the styli shorted with each other and the printing became garbled. It only took a minute or two before the styli shorted and caused the printer to malfunction. I hired Misch as a staff engineer reporting directly to me and assigned him to the High Speed Printer (he was a mechanical engineer). After he understood the problem, he suggested that we continually blow air across the styli. I agreed and he designed the hardware to blow cool air on the styli. About this time, George Shaw and Homer Denius, Radiation founders showed up in my office one afternoon. They walked in unannounced. Neither George nor Homer had ever been in my office previously. The High Speed Printer was an albatross around the company's neck and they were terribly worried. The printer development contract was fixed price and they had nearly concluded that we should ask our customer to terminate the contract. If that happened, Radiation would lose about $300,000 already spent. I had only had the project about a week and believed that we might solve the problem. Just give me a couple more weeks. They decided to go along with me. I don't think they were reassured but they were willing to give me a chance. Of course, I could have agreed with them and nobody would have thought the worse of me. Asking for more time somewhat placed the monkey on my back. Mischa Kantor's air-cooling did a lot of good in the printer performance, but it still was not good enough. The printer would run for about fifteen minutes before carbon would clog the styli. This was almost ten times better than before. Then Misch came up with another idea. He suggested that we inject ground walnut shells in the air-stream every minute or so. Mish said that walnut shells were used to clean jet engines. They were injected while the jet engine rotated at high speed. Walnut shells were abrasive enough to knock carbon off engine blades, but they were not hard enough to abrade bearings like sand would if particles got inside. Misha built the equipment to inject walnut shells and it worked great. Between the cooling air and shells, the High Speed Printer worked as specified and had a reasonable lifetime. We delivered it to the customer and, as far as I know, he is still using it for atomic bomb development. I don't remember anything else that Mischa Kantor did for us, but this project made his hiring a brilliant move on my part.

Many good friends still live in Melbourne. Though we moved to San Antonio in 1973, twenty-five years ago, we still see them sporadically. My fondest memories are of the working level grunts I knew at Radiation: Mel Cox, A.B. Amis, Roland Moseley, and Guy Pelchat. I first knew Cox at Mercer University. He arrived about the time I left forPre-Midshipmen School. He was already at Radiation when I connected with the company. Once Mel left Radiation to become Chief Engineer at another company and Mel divorced after we left Melbourne, and he married a second time. His second wife died a few years ago. The last time I saw Mel I learned he has prostate cancer and emphysema. But he still smokes. (Now in 1998 he has quit.)

As working Radiation engineers, my crowd enjoyed intellectual subjects, like: "How can light be a wave and simultaneously a mass-less photon?" Or Cox challenging me to see the planet Venus during daylight. After Sol and Luna, Venus is the brightest object in the sky. When we see it during dark hours, it is either the "morning star" or the "evening star." It never passes overhead during darkness. Like Mercury—difficult to see even at night—Venus orbits inside the Earth. All other Solar system planets are outside Earth orbit and they can pass overhead during nighttime.

Mercury and Venus only pass overhead during daylight hours. Thus Mel's challenge. He stared up into the noon sky and after a few minutes said, "There it is!" He pointed. I looked and saw nothing but blue sky. But knowing Mel Cox, I believed he found Venus. He said so. Mel told me, "You just need to know where it is. At this time, I know it is close to our meridian. Since you don't have any stars to focus on, you have to learn to focus your eyes to infinity." I looked where he pointed and tried to focus my eyes as instructed. Focusing is hard to do when you are looking overhead and there is nothing to focus on. I didn't find Venus that day. After two or three days of trying, with Mel's help, I finally saw Venus in broad daylight.

He calls me "Hoss." A.B. Amis, III has been my friend ever since I went to Radiation in 1957. He was a working engineer when I arrived. The letters don't represent anything—A.B. Is his name. His e-mail address is AONLYBONLY@MSN.COM. A couple of years ago A.B. and wife Fran entertained at their home in Grant, Florida just south of Melbourne. The party was a gathering of Radiation "Old Timers." The Amis' built their new home after Sarah and I left Melbourne. It was a pleasure to see their country place. The party was outstanding and we renewed old acquaintances with a vengeance.

I'll never forget a discussion I had with A.B. Years ago. It was a boss to employee discussion. A.B. Never showed any interest in being a manager—he always worked as an individual contributor. I felt that A.B. had leadership qualities that would make him a good manager—he certainly had the intelligence to perform well. Our conversation was my attempt to motivate A.B. to become a manager. He said he didn't want to be in management. He was happy doing his thing. I pressed the point. Finally, A.B. said, "Hoss, are you happier than I am?"

Occasionally I detected clues that led me to believe that some employees were not very busy. I recall an experience at Radiation like that. When I was running the RF Systems Department, I developed a routine I called the "Rule of Foot." The rule was that I should get off my buns and find out what was happening. To this end, unannounced I visited work areas, offices and laboratories. I wanted to see the people and their work. Thus, I showed my interest and helped with advice as appropriate. Once when exercising the "Rule of Foot," I visited Mel Cox's office. While we chatted, I noticed a new picture on his wall. It was a photograph of me, changed to give me a weird look. Someone had cut an 8" X 10" photograph across my forehead following my wrinkle line. They removed about one-third of my forehead and put the remaining pieces together. The ears were inverted and placed in opposite locations. It was a picture of a strange but still recognizable Harold O'Kelley. Not knowing what was going on, I didn't let on that I noticed the picture. After a time, I left without commenting about the image. Next I was in A.B. Amis' office. Again on the wall he had a hacked up photograph of me. It was completely different. My eyes were upside down so I appeared to be closely examining the cleavage of a sweet young thing. Altogether, on this "Rule of Foot" excursion I ran into seven or eight different pictures of myself. Later at some party, Mel and A.B. gave the pictures to Sarah. She still has them somewhere. They are funny! However, Mel and A.B. needed more work. I accommodated them.

We counted Roland and Marita Moseley among our good friends. I called her "Dandy" (a bread brand named "Meritta" reminded me of Dandy Bread—both were bakery goods sold in Melbourne). Sadly Dandy passed away undergoing surgery to clear her carotid arteries about a year ago. Roland was an engineer specializing in microwave engineering. Antennas and antenna-feeds were his forte. Roland was an Engineering Section Head responsible for developing and building antenna products. To prove his designs, Roland operated our antenna test range. One time Moseley felt his antenna range needed improvement. The changes he wanted required a capital investment of more than fifty thousand dollars. For an expenditure that big, he needed the approval of the Systems Division General Manager—me. He briefed me on the project. Moseley brought along some of his cohorts—Mel and A.B. were there. The briefing was held in my conference room where Roland chalked details of his proposal on my brown-board. After listening to his briefing, I asked if he had considered alternatives. Roland was not prepared for that, so I asked him to come back with three approaches. A few days later Roland was back with alternatives. He sketched the three on my conference room brownboard. Then he briefed me and identified his preference. "Roland," I said, "toss me a piece of chalk, please." As I caught the chalk, I told Roland, "I'm going to throw this chalk at the board. The alternative that is closest to the chalk mark is the one we will implement." As mouths gaped, I threw the chalk. I hit the alternative Roland preferred. He got what he wanted. The troops never knew that if I missed Roland's choice, I would have tried again.

Guy Pelchat is a native Canadian and a graduate of that country's McGill University. "Ghee"—French pronunciation of "Guy"—was an engineer and an able mathematician. Most of his work for Radiation was analytical. Guy graciously dedicated one of his published papers to me. As a manager I regularly held staff meetings. I particularly recall a meeting I held while I was Director of RF Systems. Guy Pelchat attended, not as a manager but as a senior technical contributor. The meeting was in my office. My furniture was arranged in a "T" shape with my desk at the head of a conference table. The meeting was held at the end of our work day. After we completed our meeting agenda, I said, "That's it. Let's go home." Most of the participants left but the usual gang hung back and placed their feet on the table—actually their feet were already up. I joined the feet up posture and a bull session began. Mel Cox and A.B. Amis were there and so was Roland Moseley. As usual, during the meeting Pelchat hung back from the table and sat in a corner. However, for the bull session he joined the group at the table. After we explored various subjects, A.B. asked me, "Harold why do you work so hard? You're always working. Why don't you slow down?" "I enjoy my work," I responded. A.B. pressed the point. "Don't you need some variety? Like doing something for fun?" I said, "I enjoy working. I get my kicks out of working." Quietly Pelchat spoke up. In doing so he broke us up. "Harold, either I don't know how to work or you don't know how to f--k."

When I was General Manager of the Systems Division, all of my financial help came from Radiation's Corporate Controller, Bill Kercher. Not only did Kercher do our accounting and run balance sheets, he also had responsibility for data processing. At first IBM punch card machines, sorters, collators and the like were used for our data processing. Next we graduated to an IBM 1620 electronic computer which was more of a scientifically oriented computer than a business data processor. Then along came the IBM 360 series. We started with a Model 30 and then had to bring in a more powerful Model 50. The data processing staff was not very responsive to users. They treated their users as though data processing was black magic. If an engineer needed a computation to be made, he had to place the order at the data processing window. Then he would be given a delivery date, probably two weeks in the future for something that should take no more than two hours. We in operations were continually complaining about poor service. As our computers became more and more technically complex, the unwieldiness of the situation became most burdensome. Finally, Joe Boyd had enough and decided to make a change. As if I didn't have enough to do as General Manager, Systems Division, Joe assigned data processing to me. He said, "Fix it!"

As part of my education, I sent myself to a one-week IBM seminar titled "Executive Computer Concepts," held at the IBM Homestead, Endicott, NY. The seminar was quite beneficial. We learned about the principles of programming in different languages, data processing technology and organization. The experience showed me that IBM tried to be involved in the hiring of every corporate data processing manager. I didn't want another corporate politician as my data processing manager. When I got home, I moved an operations manager, the Director of Reliability, Jim Robertson to the top position in data processing. This turned out to be an excellent move. The organization started acting as though they were a service group for the first time. Jim succeeded handsomely in this role.

"Robey Green and his Flying Machine." That's how we described our company pilot. Radiation had owned an airplane ever since I joined the company. George Shaw believed that the convenience of a private airplane improved the efficiency of key personnel. Thus the company got additional working hours from people who couldn't be duplicated. Our company plane was a two engine Aero-Commander when I joined the company. Shaw believed that the safety benefit of two engines offset the additional cost. Robey was qualified in many different propeller aircraft, and qualified as a flight instructor and as a master aircraft mechanic. These attainments were the reason that our aircraft operation ran at such low cost. When he wasn't flying, Robey spent much time on maintenance. Usually Robey flew non-stop for a trip within range of the aircraft. Frequently we flew to Dayton, Ohio—Wright Patterson Air Force Base is located there—which took about six hours, one-way. Since the aircraft had no bathroom facilities, such flights were an extended test of bladder flexibility. After a few years, Robey was permitted to purchase weather radar for installation in the Commander. Before that, I was on a flight from Melbourne to Dayton. We left late in the afternoon with a full load. In addition to Pilot Robey, there were five of us on-board. Somewhere after dark, maybe we were over Tennessee or Kentucky, we ran into some serious weather. The Commander had automatic audible warnings. If the pilot pulled the throttle back to idle and the wheels were still up, a horn sounded in the cockpit. This was to warn of a wheels-up landing. In the bad weather we encountered going to Dayton, we were bouncing around, falling for a thousand feet or so and being pushed down in our seats as the aircraft was swept higher in updrafts. Frequently the warning horn sounded—Robey pulled the throttle back to idle to reduce our rate of climb in updrafts— which added to our fright. After a seeming eternity, we got through the weather. When we landed I was surprised that I still needed to empty my bladder.

When small corporate jets came onto the market, Shaw made sure we were among the earliest companies to own one. Radiation purchased a new Learjet-23. An indication as to how early in corporate jet purchases we were, our aircraft was Serial No. 11. Robey had to go to jet flying school. We hired additional pilots since the Lear required a crew of two. Usually we flew the jet with a full load of six passengers plus a crew of two. During the time I was Senior Vice President of Radiation and Vice President-Programs for Harris—a staff position I became supervisor of the aircraft operation. Green reported to me. On a flight to Quincy, Illinois we had our first accident. (I wasn't on board.) The flight crew consisted of two ex-military pilots. The captain had over 5,000 hours of jet time. The other pilot was not jet qualified but he had 12,000 hours of propeller flight time. Unfortunately, the captain was in the right seat and the non-qualified pilot landed the Learjet. It had a full load. The runway at Quincy was being extended and there was only a rock base on the extension. The old part of the runway had been used many times by our crew. The pilot landed long and the jet-qualified pilot, sitting in the right seat, had his head in the cockpit and wasn't watching outside. The result was that our Learjet went off the end of the runway shearing off the landing gear and causing scrape damage to the undercarriage. Thankfully the passengers and crew were only shaken up. There were no injuries, but the aircraft was severely damaged.

When Robey called me and told me of the accident, I asked him to see that the aircraft was not moved until the FAA could examine it. This kept the runway closed for a few days. I visited my boss, Joe Boyd, and told him of the accident. The accident kept me busy for a few days. I leased another Learjet for company use. The leased aircraft was a Learjet-24—longer and carried more passengers. We fired the two pilots. Lear engineers went to Quincy and made temporary repairs. With special permission from the FAA, Lear pilots flew our plane to their plant in Wichita for permanent repairs. They had to fly at lower altitudes than the normal cruising altitude of 41,000 feet. Because its flight time approached 5,000 hours, our Learjet was inspected using the 5,000-hour routine. We flew our aircraft more than most and ours was the first Lear-23 that went through the 5,000-hour inspection. That process included removing the wing sheet metal to look for corrosion.

I learned much about flying from Robey. Usually when I flew with him I sat behind him and observed and asked questions. On a few occasions I sat in the right seat of the Aero-Commander and steered the plane. I never tried to steer from the right seat of the Learjet—it was too sensitive for me.

Radiation had an especially good Industrial Relation Department headed by Curt Gallagher. Key people in the department included Jack Wilson and Bob Underill. Jack and Marilyn Wilson were our personal friends. (Jack teased Marilyn about her large mouth, "She can bite a watermelon," he said.) Before Radiation, Jack played on the Ohio State football team. He was team captain one year when they went to the Rose Bowl. I asked Jack why he didn't play professionally and he responded, "I couldn't take the cut in pay." When Jack visited our home, he wrote his initials in the dust atop our refrigerator. This wasn't a problem since only he was tall enough to see it. Sometimes we went on weekend excursions with the Wilsons. Once Elting and Julie Storms, the Wilsons and Sarah and I spent a long weekend at the Doral

Resort in Miami. I was a catastrophe playing golf with the men. I drove my ball into the rough and while searching cut my forehead on a palm frond. And when I was putting on the last hole, my glasses fell to the ground. Jack and Elting complained that I ruined their game from watching me. Bob Underill was very effective in dealing with employees. Bob's brother, Harry, was administrator of the local hospital, and an excellent artist. He paints in oils. Harry attended Mercer in the Navy V-5 program at the same time I was there. Bob worked directly for me when I was Acting Manager of the Control Division, and Harry retired and moved to Colorado where he paints full-time under the name "Hank Underill."

One weekend, we joined in a fun trip to Miami. In addition to Sarah and I, party people were Jack and Martha Hartley and Jim and Judy Brantly. Hartley was a Radiation marketer and had been a student of mine at Auburn. Brantly was a scientist and had a doctorate. Jim headed the Radiation Research Division in Orlando. My friendship with Brantly resulted from our working together on proposals. Our group stayed at a motel on the beach. We followed the usual beachside activities with visit to nightclubs taking up the evening hours. The other couples left a day before we did. The last night in Miami, Sarah and I were down to our last twenty dollars. In those days we had no credit cards, and we were strapped for money to pay for our final dinner. We went to a restaurant called "Place for Steak." We were seated against the wall at a small table for two. Sarah had her back to the wall on a bench that was attached to the wall. The bench extended to include two or three tables either side of ours. I was seated in a chair across from her. Sarah had on a dress that she had personally made. A lady seated next to Sarah wanted to know where Sarah bought her dress and that made her evening. We had a wonderful meal and paid the bill. Obviously we ordered carefully.

At one time I supervised the Radiation operation located at Malabar airfield. The facility had been built during World War II as an adjunct to the Naval Air Station at Banana River. It had an asphalt runway and a small frame building stood close by. We operated military radars at the site for testing for aircraft evasive countermeasures. Chaff is made from small strips of aluminum foil that are packaged so that, when thrown from an aircraft, they scatter into a small cloud and slowly drift to earth. If unfriendly fire-control radar locks onto an aircraft, frequently a chaff cloud can spoof it. The chaff returns a signal larger than the signal from the aircraft. A tracking-radar interprets the two returns and follows the larger signal. Thus a break-lock occurs at the tracking radar and anti-aircraft shooting accuracy is impaired. A countermeasure to the chaff is for the tracker to follow the higher velocity target. This can be achieved using the Doppler effect. Thus, radar lock-on could be maintained. Electronic warfare becomes an endless application of a countermeasure to a countermeasure to a countermeasure. No single countermeasure is effective for long. One technique developed by the Air Force with Radiation's help was to dispense chaff to the front of the aircraft. Radiation built a teardrop shaped pod attached to an aircraft's exterior. From the pod, rockets were launched forward of an aircraft when radar lock-on was detected. An explosive device in the rocket caused a chaff cloud to disperse. This cloud was in front of an aircraft instead of falling from the rear. This countermeasure was effective for a few years but other countermeasure techniques soon caught up.

The purpose of the Malabar site was to measure backscatter from chaff clouds dispensed from real aircraft. The chaff was dropped over the Atlantic, but sometimes we received complaints from cattlemen about their cattle eating aluminum foil. The chaff drifted for long distances in the winds aloft. Therefore, we tried to conduct our tests when winds favored our efforts, but mostly there was an onshore breeze. We also tested forward launched chaff. That meant that rockets were fired directly at us since the aircraft usually flew directly on-shore from the Atlantic. However, we gave the order to fire the rockets when the targets were well out to sea. On one occasion we tested the radar back-scatter from a T-33 jet trainer. The airplane had been covered with a radar absorbing material and the objective was to make the aircraft less visible to radar. Thus we were in on the ground floor of the "Stealth" technology that has resulted in military aircraft that are almost invisible to ground based radar.

Our second car was a little Fiat. (Before that it was an old Plymouth coupe that had a big hole in the floorboard. I paid $75 for it.) The Fiat was smaller than a VW and it had an air-cooled rear engine. The car was grossly under-powered. Winds from the Atlantic to shore prevailed. The Malabar field was west of Melbourne. When I traveled from my Melbourne office to Malabar, invariably I had a tail wind. On the highway I floored my Fiat. The speed differential going and coming was about ten miles per hour. I traveled to at about sixty and my return was at fifty. Once when I was at Malabar reviewing project status, when I left my Fiat wasn't where I parked it. I looked all over and finally I found it in an office. It wasn't driven there. With help my good friend and disrespectful subordinate, Guy Pelchat, had picked-up my car and moved it in there! After a good laugh, Guy carried my toy back outside.

The U.S. Space Program had much difficulty in its early stages. Our attempts to launch what the USSR leaders described as a "grapefruit" sized satellite failed because of underpowered rockets. Our rockets showed a damnable tendency to explode at or near launch. NASA's publicly viewed launches invited criticism from the news media as well as from foreign competitors, especially the Soviet Union. After President Kennedy issued a national challenge of putting a man on the moon NASA established the manned space program. Considering NASA's lackluster record for satellite launches, the President took a risk in assigning them as the primary agency to meet his goal of placing man in orbit and venturing farther to the moon by the end of the 1960s decade. Thus, NASA undertook a step-by-step methodology of space experimentation. They achieved the President's goal when Neil Armstrong, on July 20, 1969, stepped on the moon and said, "That's one small step for man, and one giant leap for mankind." The U.S. Military establishment was envious of NASA. Pentagon leaders thought that spy and communications satellites would bring military advantages to America. Especially they felt that a manned spacecraft had military potential. It wasn't well known at the time, but the military did have a manned spacecraft program. The Air Force undertook a project to build an aircraft that would fly into space and return. The Air Force named this project Dyna-Soar. My engineering organization at Radiation won a contract related Dyna-Soar. Jointly, RCA and Radiation bid on the electronics for Dyna-Soar and we won the contract for the spacecraft electronics and ground electronics. For Radiation I had the responsibility for the ground tracking antennas. Not long after we won Dyna-Soar, the Air Force abandoned the project, and canceled our contract. We never built anything for Dyna-Soar.

The first manned space program had a goal of putting a lone man into near earth orbit. NASA named the project Mercury, because Mercury is the first planet in our solar system. The first United States Astronaut to enter space was Allen Shepard, Jr. His was a sub-orbital flight. It was comparable to shooting a missile at a target on the earth. In fact, Shepard's flight was that of a manned missile. Shepard made his flight in May 1961. John Glenn, Jr. (later U.S. Senator Glenn, D-Ohio) was the first U.S. Astronaut to orbit the earth. (Little Shannon said, "I just wove John Gwenn.") His pioneering flight was on February 20, 1962. His one-man Mercury space capsule circled the Earth three times in about 5 hours. There were three more Mercury manned orbital flights before the two-man Gemini spacecraft was ready with a booster powerful enough to lift it into orbit. Gemini was the world's first maneuverable spacecraft. Before a manned flight took place, there were two unmanned Gemini flights. Gemini 3 was the first manned Gemini flight. It took place in March 1965. Gus Grissom and John Young were the astronauts on board for that Gemini first. During the flight of Gemini 4 (June 1965) Ed White became the first man to make a space walk. In the last month of 1965, two Gemini capsules orbited and were in space simultaneously. Gemini 6 and Gemini 7 were the first spacecraft to maneuver into docking positions. They were within one foot of each other while orbiting the earth.

My contribution to the Mercury and Gemini programs was to build tracking antennas for communications and for telemetry acquisition. We had to build the antennas in a very short time since the space program was going all out and we were always in a hurry. President Kennedy's national goal created a sense of urgency. We made the antennas from surplus Army searchlight mounts and a Radiation built aluminum antenna structure with automatic tracking electronics. The antenna was a flat metal plane, on which we mounted four parallel helical devices made from tubing. The use of four antennas made for stronger signals, and gave us four signals that could we processed to make the antenna automatically track a target in space. The Mercury and Gemini projects were successful and so were the improvised antennas my engineers built.

In the 1960s, neither the U.S. Military nor NASA had rockets with enough thrust to put a payload in geo-synchronous orbit. (Geo-synchronous orbit is one that has a satellite revolving around the earth and making one revolution while the earth turns once. Thus, from earth the direction to the satellite appears to be constant.) The U.S. Army wished to take advantage of space born communications devices. At first they settled on using a low altitude satellite as a courier. A ground station would dispatch a message to the satellite. There a tape recorder stored the message. When the satellite came into view of the addressee station, a ground station would command the satellite recorder to play back the message. Project Courier was a sophisticated version of the first repeater communications satellite known as Project Score. Score was the satellite that broadcast President Eisenhower's holiday message in 1958.

The Courier satellite was spherical in shape and had a 51-inch diameter. Crammed into the sphere weres separate electronic packages including tape recorders, transmitters and receivers, tracking beacons, telemetry equipment and battery-solar cell power supplies. Courier's mission was to prove the feasibility of active satellite communications. It operated in two modes, (1) A delayed repeater or "mail" pouch, and (2) a real time repeater. To conserve satellite power, the tape recorders read out in reverse, thus obviating the need for tape rewind. My engineering section worked on Courier in that we designed and built tracking antennas used on the ground. These antennas were 28' foot diameter paraboloids that could automatically track the satellite transmission as the bird traversed a line of sight. Radiation contracted to build three of the antennas: one was for an Army installation in New Jersey, and second, was close to Madrid. I forget where the third was to go. Early in the project I personally performed some to the system studies. It was during this study that I realized a geographical fact that at first seemed wrong: Madrid is at about the same north latitude as New York City. Of course, this means that London, Paris and Bonn are far north of New York.

I suppose booster technology advanced more rapidly than expected. Courier proved the feasibility of "mail pouch" satellite communications, but the technology was never placed into military operation. The Army canceled one of the three ground stations. The need no longer existed. Later the Mark V ground stations that we built were operational.

Manned space programs have advanced since the days of the Apollo mission to the moon. As I write, the Space Shuttle is in orbit and the Mars Pathfinder sits on that planet. Rover roams the red planet. MIR, the Russian space station, orbits with danger lurking on the horizon.

A Vietnam War era classified program that Radiation was heavily involved in had the code name "Igloo White." When reading *Compromised: Clinton, Bush and the CIA*, by Terry Reed and John Cummings I was surprised to encounter a description of Igloo White in the first ten pages. The Igloo White system was based on sensors dropped from airplanes in the Demilitarized Zone, or on the Ho Chi Minh Trail, or in Cambodia or Laos. Location was determined by the need for intelligence about different areas. Depending upon what was being monitored, different sensors were used. Some of them were seismic (listening for large vehicles, tanks and the like), audio (listening for troops voices and marching feet), and infrared (for nighttime detection). The devices were dropped from airplanes and consisted of a stick with a spike on the end having the heaviest weight. When the spike hit the ground, it stuck in the dirt deep enough to hide the unit. It thus became an intelligence collection station in the enemy's country. Because of the shock upon striking the ground, the electronics had to be rugged. Therein, our aerospace technology came into play. Radiation's Aerospace Engineering Division was expert in designing equipment for survival in severe environments such as rocket launches, space and flying aircraft. Radiation made the data processing equipment and the radio transmitters that went into the spikes. Other companies manufactured the sensors. The relay part of the system was equipment that flew in an aircraft (a forerunner of the Air Force AWACS) circling over South Vietnam. We built the equipment that went into the relay aircraft. The system receiver and processor were located near Bangkok, Thailand. Radiation built that ground based equipment and operated it for the Defense Department. My friend, Mel Cox, went to Bangkok as the Program Manager. At the time, this project was highly classified but Mel tells me it now has been declassified. This system has been described in the newspapers as McNamara's (Secretary of Defense) Electronic Fence. I have read of an electronic fence used in the Sinai Desert to keep the Egyptians and Israelis separated: I suspect that it was the same McNamara's Electronic Fence technology.

When I joined the RF Division, Bill Beatty was a staff engineer and a systems analyst, with no desire to be a manager. Bill was well respected for his creativity and his ability to put his ideas across. Perhaps, Beatty was ten years my senior. After I became an engineering manager I felt that Bill looked down his nose at me. When we were traveling to San Francisco for a customer visit, one evening Bill and I went to dinner. He took me to a Japanese restaurant, Yamato's. This was my initiation to Japanese cuisine. Yamato's was an authentic Japanese restaurant. We had to remove our shoes and sit on the floor, but thankfully, there was a well where our feet went. Beatty taught me to use chopsticks. Some years later, Beatty left Radiation and went to Philco, Western Defense Laboratories, in Palo Alto, California. Still later, management at WDL approached me about joining their team and heading up their tracking antenna department. Sarah and I went to Palo Alto for interviews, and this was her first trip to California. It surprised me to learn that Bill Beatty had recommended they recruit me. He had more respect for me than I thought. Probably, he scorned all management as a necessary evil.

In March 1960, management shook-up Radiation's Melbourne operation. The company had hired an engineering manager from ITT-Kellogg, Dick Hultberg. He was appointed Director of the new Systems Development Division, Melbourne. Under Hultberg were two departments, Data Systems headed by Mel Cox and RF Systems with me as the manager. The RF Systems Department was the RF Division renamed. After we returned from Palo Alto, before Philco had made me a firm offer (our considering leaving was a secret) Eric Isbister, Vice President of Engineering, asked me to take over the Data Systems Division as he was relieving the Director, Dick Hultberg. Hultberg had not fit well into the entrepreneurial spirit of Radiation. He tended to be a pipe-smoking bureaucrat. (Data Systems Division had responsibility for airborne telemetry equipment and ground based data processing equipment. I ran the remaining engineering organization, RF Division.) I told Eric that I couldn't accept the Data Systems position because I was waiting for an offer from Philco. I told him about the job and that I had told them that I wouldn't consider the job for less than $25,000 per year—I was making $19,800. Before the day was out, Isbister called me to his office and told me I would receive a raise if I accepted the new position. My new salary would be $25,000. I accepted.

Eric Isbister was from Sperry. As Radiation Vice President-Operations he was responsible for Engineering, Program Management, Manufacturing and Quality Control. Eric was strange. He was a chubby fellow who chewed cigars and, I particularly recall when he ate fried shrimp, he consumed tail and all. He had a tough job and had to resolve differences between organizations that were headed by strong-minded individuals. I was one. As time went by, Eric got into more and more difficulty. He often made a decision regarding an issue existing between two organizations by responding to the push of one person. He didn't make sure that he knew both sides of an issue. When his subordinates came to understand this, a corporate political battle started that caused Eric to procrastinate. He couldn't bring himself to make decisions. In one instance, I visited Isbister in his office and, in private, I lectured him on his problem. That's a strange thing—a subordinate chastising his boss. As I talked, Isbister just looked at me with a passive expression on his face. He didn't show any anger though he should have. I left his office dumbfounded and concerned that I had to work for a man with such little backbone. A few weeks after my private session with Isbister, Joe Boyd called me to his office and promoted me to Isbister's job. I would be the new Vice President of Operations. Eric was placed on staff until he could find another job. My promotion to Vice President was special. When I arrived home that evening Sarah was as excited as I. I blurted out the news before getting out of the garage. Jack Hartley came by our house and congratulated us. I thought that was gracious of him, especially when I later found out he wanted the job—he was Vice President, Marketing. As VP Operations I was responsible for all engineering, manufacturing, quality control and program management for our government business.

Later—April 1964—I was promoted to Vice President, Manager of Radiation Melbourne. In addition to Operations, Marketing, Contracts and Purchasing, Advanced Communications, and Facilities were placed under me. I had the functions of a complete business except for financial, which was provided by Bill Kercher reporting directly to our President. Dr. Al Sissom was appointed to my old job. In 1965 I was given the title VP-General Manager, Radiation Systems Division. I then had a financial controller. From 1957 until 1965, I had moved from Project Engineer to Division General Manager. Like Rodney Dangerfield, "I ain't got no respect." A.B. Amis said, "Hoss, when are you going to learn to hold on to a job for a decent interval?"

The male voice on the military air transport public address system boomed, " Your attention please. The Commanding Officer requests that all personnel wait until Mr. O'Kelley and his associates leave the aircraft." I couldn't believe my ears. This was embarrassing. Jack Davis told me this would happen. He had made the long trip before and should know. He also told me that an Air Force Band would be playing martial music at my arrival. "This is standard Air Force treatment," Davis said. When he told me this before, I had not believed him. Jack is a practical joker. However, at Hickam Field in Hawaii a young First Lieutenant escorted me aboard a DC-6, Air Force version. My traveling companions were Radiation people, Jack Davis and Bill Corry. We were on our final leg of a tiring journey from Melbourne to Kwajalein Atoll, Marshall Islands, in the Pacific Ocean. Kwajalein is about 2,500 miles west-southwest of Honolulu. Flying time on the four-engine propeller driven aircraft was about ten hours. " Kwaj" is at the same latitude as the Panama Canal. We did not have to cross the equator to get there, but we crossed the international dateline.

It was late December 1967. Davis was our Program Manager for Project Altair and Corry was a key mechanical engineer on the project. Jack was a young, aggressive electrical engineer with experience on large antenna projects. He once moved his family to Ethiopia when he was in charge of an 85- foot antenna project. The National Security Agency used this system as a listening post. Jack's daughter and our Shannon were good friends in grade school. When Jack and Betty discovered that their daughter had leukemia, Jack told me he intended to resign. They would move to a locale where she could get better treatment. I encouraged him to stay feeling that he could get adequate treatment for her where he was. He decided to stay with Radiation and moved up through the ranks. I helped him in one of his moves. At my insistence we transferred him to manage engineering in the Control Division when I was there temporarily. This move started him on a fast track of promotions. A few years after I went to Datapoint, I tried to recruit Jack Davis to join me as President and top operations man. I would remain as Chairman and CEO. The offer was tempting to him, but finally he refused. Davis thought that he had a good chance to become the next President of Harris when Jack Hartley relinquished the title. He was right at the time. Something happened. He was Sector Vice President of Harris when he left. Davis became Chief Executive Officer of Data Products, Inc., a printer company with headquarters in California. Later, when an unfriendly takeover occurred at Data Products, Jack left. He came out very well financially and today he has a 600-acre farm in Mississippi and raises horses.

Bill Corry was a top mechanical engineer, and he came from a family that was Radiation to the core. His brother, Doug, and Doug's wife, Mary, worked for the company. Our company and Sylvania had teamed to bid on the Altair Radar Project. Altair was a code name. Sylvania was the prime contractor and we were subcontractors. Radiation designed and built the 150-foot diameter dish antenna. The structure was so heavy that the Altair antenna azimuth bearing consisted of railroad wheels running on a circular track. The purpose of the dual frequency radar was to gather radar signatures created by ICBM warheads approaching impact. The Air Force launched missiles at Pt. Mugu, California and their warheads landed in the Kwajalein Lagoon. The radar with its large antenna would track incoming warheads through reentry and splash down in the lagoon. The three of us were ending our trip to visit Pres Shrader and his crew of antenna installers at Roi-Namur. They lived in military quarters on the main island, Kwajalein, and flew to work about five miles away. The Air Force provided shuttle service. As visiting firemen, we stayed at bachelor officers' quarters and took the same shuttle flights to visit the site. We were civilians in military space and they treated us as officers. Our Air Force travel orders made us into Civil Service equivalents. Corry was equivalent to a GS-14, Davis GS-15, and I was a GS-17. My rank was the civilian equivalent of a general. Military people treated me as though I was wearing stars on my collar. Thus, everyone had to wait my deplaning since I was the ranking officer aboard. Jack was kidding about the Air Force band, but everyone waited until I got off the airplane. The station protocol officer greeted me at the bottom of the steps. We visited with Pres and other Radiation people. It pleased them that we made the effort to travel so far to see them and their work. It was a good morale building gesture, but it accomplished little else. Corry and I considered it a bit of a treat but Davis had already made the trip more than once so he wasn't all that thrilled.

Radiation had a contract with the U. S. Navy to develop an airborne flight path guidance system for carrier landings. At the time I was running the Systems Division. The purpose of the ASW-25 was to interpret the glide path toward an aircraft carrier as received over radio waves. It acted much like the glide slope installed at many commercial airports. The ASW-25 was state-of-the-art equipment using microelectronic circuits. The Radiation Microelectronics Division under Jack Hartley wished to be the microelectronic supplier for the circuits. My Systems Division people, and me, responsible to the Navy didn't want to take a chance with the Radiation Microelectronics Division because the division was young and untested. This was responsible for the worst management conflict I ever had with Jack Hartley, and ultimately with Joe Boyd, our boss. As Radiation President Joe Boyd had to make the decision as to where my Systems Division would buy its microelectronics parts. He decided in favor of Jack's division. He was right. Though there were some difficulties along the way, the Radiation Microelectronics Division came through on the ASW-25. I'm glad that I didn't have other conflicts like this.

We had a contract with British Aircraft Corporation, one of two developers of the Concorde supersonic aircraft. Their partner was the French national aerospace company. Radiation developed digital telemetry for the prototype aircraft flight test. The French were developing analog telemetry equipment for use in their tests. Even though it was for flight, the Data Systems Division as opposed to our Aerospace Division was performing the work. The equipment was rack mounted in Concord's passenger cabin and was not designed for rigorous flight environment. Radiation also built the telemetry data processing equipment to evaluate the data. As General Manager of the Systems Division I was the responsible senior executive. Once while traveling in Europe visiting Intertechnique, I made arrangements to visit British Aircraft along with Dub Hudson, Radiation staff legal counsel. The BAC plant is in Cardiff, Wales. We were staying in London and BAC sent a car for us. The ride to Cardiff was through some beautiful countryside. We were treated as VIPs. BAC management pulled together about a dozen people to meet with us. After lunch we were shown the full-scale mock-up of the Concorde. I was surprised at the small interior. Engineers explained that to efficiently cruise at supersonic speed the ratio of fuselage length to diameter was constant. Increasing cabin size (fuselage diameter) required lengthening the aircraft fuselage. Increasing these dimensions resulted in a weight increase. Thus, engine power requirements would go up. The tradeoff resulted in a small cabin. In later years when I was at Datapoint, I flew the Concorde across the Atlantic on three or four occasions. I traveled both with British Airways and Air France. The trip from New York to London and from New York to Paris was about the same—three hours. It wasn't nearly as comfortable as a wide body jet like a Boeing 747 or McDonald DC-10. I much prefer a wide body when it comes to comfort. Though it takes more than twice as long, amenities are far superior on the wide bodies. And one can fly from Miami or Atlanta eliminating the bother of getting to New York's JFK airport.

It had been a quick turnaround—an over-nighter for Bill Premaza and me. The previous afternoon, we flew from Melbourne to Atlanta, and then to Newark where Dom Casale met us. (At that time one had to go through Atlanta to get to heaven.) Dominick drove us to Fort Monmouth in New Jersey for a next day meeting. Our host for a ceremonial contract signing was Colonel Mitchell Goldenthal, Commander, U.S. Army Satellite Communications Agency. Mitch was a bird colonel. Though I was Vice President, Operations I deeply involved myself in the Mark V proposal preparation and marketing. I came up through the engineering ranks where we designed tracking ground antennas. The Mark V was a "mil spec" satellite ground terminal operating at X-band. The contract called for design, development, and manufacture of ten terminals. It worked with low orbit satellites (about 100-200 miles). This required that the ground antenna automatically track satellites as they moved through the sky. Line of sight to today's satellites with high orbits (about 23,000 miles) appears constant because their orbital period equals that of one earth rotation. Thus, satellite antennas today look at one place in the sky and do not need to track. The Mark V tracked satellites that had low power output. The antenna system was transportable on the earth's surface or in aircraft since the antenna structure folded up. We designed the antenna and electronics trailer for battlefield conditions. These included rough terrain and extremes of temperature, altitude and humidity. Our New Jersey trip was a culmination of a lot of work. The contract was for about five million dollars and it was one of the largest the company had won until that time (1965). We were proud of beating companies like RCA, ITT, Sylvania, etc. Radiation, Inc. won against the big guys with household names. The signing went as planned. Afterward, Dominick drove Premaza and me back to Newark where we caught a plane to Atlanta, continuing to Orlando. At Atlanta, I met a Radiation Customer from Grumman Aircraft. (Under subcontract to Grumman, Radiation was developing the digital telemetry equipment for the Apollo Lunar Excursion Module). When we arrived in Orlando, he invited us to ride to Melbourne in his rental car. We accepted with the understanding that we had to go to the Melbourne Airport. Premaza said his wife left his car there. As we drove, we discussed the Apollo program. We were having difficulties with the Elgin Company (watches) in their development of a clock for our equipment. (The clock was sophisticated electronic equipment that established fundamental timing for the spacecraft.) Approaching the terminal building in Melbourne, I realized there was a crowd in front. It was a Radiation welcoming committee. My Mark V Team had intended meeting our plane from Orlando. Premaza called from there and told them we were coming by car. It was a loud raucous affair in front of the terminal. There was a band and microphones and amplifiers. A. B. Amis had a cigar in his mouth and a snare drum strapped over his shoulder. Homemade banners screamed, "Harold Brings Home the Bacon," and "Mark V A Big Team." Mel Cox (we were in the Navy together at Mercer University) would be our program manager. He was present grinning from ear to ear. Ray Penn celebrated with us. (I called him "Worms" because, when he was a Contract Administrator, he often brought me cans of them.) I made a short "Thank you" and " Congratulations to the team" talk. One picture shows me waving the contract in one hand with crossed tom-tom sticks in the other. Photographs of the event show me with short hair, receding (and receded) hairline, and dark rimmed glasses. Though my scrapbook pictures don't show it, Jack Robbins was there. I don't need pictures to recall his presence. I found out recently that the group had dinner at the Melbourne Beach Steak House and Joe Boyd was there. It was there that Robbins got plastered. He was in his cups and showed his joy by getting on top of our Grumman customer's car where he did the Robbins' Stomp. He badly dented the car roof. How embarrassing! I took our Grumman customer aside, apologized, and I told him that Radiation would cover the damage. He never mentioned Robbins' Stomp later as I dealt with him.

I remember with less fondness another trip to see Colonel Mitch Goldenthal. We were well into the Mark V program when we recognized we had a cost problem on our cost plus fixed fee (CPFF) contract. As was usual on CPFF bids, our original cost estimate was optimistic and we shaved it very close. Usually technical approach and experience, not price, won CPFF contracts, but price still had to be competitive. In our technical approach we used four mechanically attached parabolic dishes, and our high efficiency feed system was unique. It was necessary for us to go back to the Satcom Agency and ask for a cost overrun. We worked up a new estimate and set a visit date with Agency people. Our intention was to discuss the cost overrun with the working troops simultaneously with me telling the Colonel in a separate session. We would break the bad news throughout the Agency simultaneously. By this time Radiation owned a Lear Jet. I reserved it for a trip to Newark. We would make it to Ft. Monmouth and back in one day. The day before we were to travel, I held a dress rehearsal and final review. Those attending included Program Manager Mel Cox plus key people from engineering, contract administration, and marketing. By lunch we had our agenda set. At take off the next morning, there would be a planeload. As the workday wore on, I became more and more concerned about our approach to the Agency. I thought about the feelings that Colonel Goldenthal would have. He would feel badly about us placing him in the position of asking for more money for this key program. Further, by our informing his people concurrent with me telling him, we made it tougher for him to manage the crisis. The more I pondered this situation, the more

I thought that our approach was wrong. Toward the end of the workday, I called Mel Cox and asked him to visit with me. I told him, "Mel, I think our approach is wrong. I think that I should go to the Agency alone and talk to Goldenthal privately. I won't have to get too specific as to details. I will tell him that the cost of the program has about doubled and give him the number. I will leave the cost estimate detail with him. A week or so later, our working level folks can meet with theirs. In the meantime, the brass can consider approaches for finding the dough." "That's a great approach, boss," Mel said as he breathed easier. But, for him, it only put off the inevitable—he had to face our Agency friends later. As word got around to other members of our team, none came forward with an argument. They didn't look forward to this trip.

Dominick met me at Newark and drove me to Ft. Monmouth. While I visited with the Colonel, Dom talked to other Agency personnel. He told them he didn't know what kept the others from flying up. All he could say was, "O'Kelley ordered them to stay home." He knew what was going on but he lied. I should not have told him until after my meeting. But he was and is a good scout and played his marketing roll with skill and finesse. Mitch Goldenthal was a tough, regular Army Signal Corps officer. He hoped for a brigadier's star someday but never made it. Possibly our cost overrun hurt his career. It didn't help. I still think that we made the right approach to the Agency.

The Mark V System was the first satellite ground station we built to military specifications. It was a successful terminal and met the requirements. Because of advancing satellite technology, it did not survive long. Stations with fixed antennas became standard when launching vehicles became available for putting satellites in geo-synchronous orbit. I was proud of the product and our Mark V Team—Mel Cox, A. B. Amis, Roland Moseley, Guy Pelchat, Bill Premaza, Ray Penn, and many more.

Probably the most unusual experience I had as a businessman was a visit initiated by a telephone call from my Radiation Manager of Security. I was General Manager of the Systems Division when Bob Rhodes called and said he had a man in his office that wanted to meet with me. I knew Bob wouldn't waste my time and I should honor his request. His office was downstairs one floor beneath mine and I asked him to come up and bring his visitor. The stranger knew a lot about me and about the company. He didn't identify whom he represented but I assumed he worked for the CIA. The conversation did not have a social segment—he immediately got down to business. The visitor said, "I know the superior technical capabilities of your company. And I know the reputation you have for designing and building quality products. Customers and competitors alike know Radiation's strength in PCM (digital) telemetry. We know that you have many people working here with military security clearances. And we have checked you out personally and know that you are of high character." Prepared to listen some more, I responded, "Why thank you." My visitor then told me that his company would like for us to do extensive work for them. He said, "It will be different than what you are used to. The company will have to build 'red-black' facilities inside your current buildings and control access in ways that were not necessary even with Top Secret military work." Basically, he said we had to create buildings within buildings. Access to the closed areas would be on a need-to-know basis. Our personnel would have to undergo polygraph testing to assure that we employed no security risks. They had already started a Special Intelligence Clearance for me based on my latest military Personnel Security Questionnaire. They did not expect to clear anyone of higher rank than me so I couldn't disclose to my bosses exactly what I was doing or whom I was doing it for. Thus began a business at Radiation that by now must have meant a few billion dollars of contracts. I think the most intimidating experience I ever had was taking a polygraph. This was a requirement for my permanent Special Intelligence Clearance (this was in addition to my Top Secret Military Clearance). The polygraph operator sat me facing a blank wall and helped me to relax. First he made a trial run followed by the real thing. I passed without any real difficulty except for the last question, "Is there anything he had not covered that was germane?" I answered "No" but he indicated something was bothering me. After some discussion he let it go. The polygraph operator was right. Even though there were no questions where my answers were lies, the fact that I been married before bothered me in answering this final question. However, I never had any problems with my clearances.

The telegram was date stamped May 2, 1969. It read:
WUO20A THA200 PDB FAX TALLAHASSEE FLO
2228P EDT
HAROLD E. ODKELLY
SENIOR VICE PRESIDENT RADIATION INC
MELBOURNE FLO
THANK YOU FOR THE SIGNIFICANT VOTE OF
CONFIDENCE YOUR ORGANIZATION HAS
GIVEN FLORIDA BY PROCEEDING WITH YOUR
EXPANSION PLANS. WE ARE DELIGHTED
TO LEARN THAT YOUR FINE ORGANIZATION
IS CONTINUING TO PROSPER AND LOOK
FORWARD TO MEETING WITH YOU SOON TO
EXPRESS OUR APPRECIATION.
CLAUDE R. KIRK JR GOVERNOR.

Although the governor didn't spell my name correctly, with this message we could claim victory in obtaining acceptable roads to the Palm Bay facility of Radiation Incorporated. I was Vice President-Operations managing all Divisions of Radiation when I started the Road Campaign. Originally, our Palm Bay facility consisted of 120 acres of land with most of it covered by buildings, parking lots, and antenna test ranges. There were only two two-lane roads serving the facility. One, Troutman Boulevard ran along the west boundary of Radiation's property. The community center of population was to the north of the Radiation campus. Thus, most of our people lived to the north and didn't use this road. The majority of our people used the front gate that opened to Palm Bay Road. This was a two lane east-west road. If, upon exiting the gate, one turned left, after driving about a mile a person reached Babcock Road. This street provided north-south passage but was only two lanes wide. When exiting and turning right—east—at Palm Bay Road, one headed in the direction of U.S. #1. This highway was four lanes wide north and south. However, to get to U.S. #1, there were two ninety-degree turns on the two-lane road. The turns led one across the railroad tracks. After arriving at U.S. #1, one encountered stop signs. There were no traffic lights. During rush hour, Palm Bay policemen directed traffic at the intersection. We faced a dilemma: Radiation needed to expand. We could not put any more people to work without constructing new buildings. Our employees couldn't commute to work in an acceptable way. Our key people were hard to find engineers and commute time was important in recruiting and keeping them. Around our facilities there was plenty of vacant land available at reasonable prices. Rezoning probably would not be a problem.

Harris' (Radiation had merged with them) corporate policy was not to dominate a community with company presence. However, as a practical matter, Radiation already dominated the Melbourne area, and policy should not prevent expansion at Melbourne. Roads limited us. Only the city, county, or state governments, or a combination of the three had the power to solve our problem. It would take taxpayer money. Therefore, we had to involve the community. We were in a classic situation. Our facilities were new. If anyone thought about our predicament, they probably concluded that they had taken care of us. The attention of community leaders was on other concerns, like recruiting new industry. One way to highlight our problem was to make governments and the publics understand our situation. If the governments didn't do something, we had no choice; we would take our growth elsewhere. To keep from appearing negative, we studied alternative plans to ease our traffic problem. A small group of inside people conducted the study. I led the group. We had to get the public to understand our situation. Thus, Warren Vergason, head of our Public Relations, was a key participant. Each day, for a week, we counted cars as they came in each gate. Further, we recorded the number of people in each car. Thus, we came up with the average number of cars coming and going in one day. Knowing the number of people we employed, we computed the average number of people per car as 1.4. There was no public transportation. Our first move was to undertake an internal program to lessen commuting aggravation by staggering starting times. Instead of one starting time at 8:00 A.M., we had five starting times beginning at 7:30 A.M. Other starting times were in fifteen-minute increments. The fifth starting time was 8:30. I assigned each department a starting time so that the number of people arriving at a given time was about uniform. It would help if we could raise the number of people per car. We started an internal publicity campaign encouraging people to car pool. We added an incentive. Cars occupied by more than one person could park in a special reserved area close to buildings. People who drove alone would have to walk further. When we made these changes, we saw some improvement due to starting times. However, the people per car didn't change. Many of our people worked late and they didn't know what time they would be leaving, and they couldn't car pool. We could say, however, that we had done what was possible, but still a bad situation existed. Improvement in roads was the only solution.

After much effort, we settled on recommendations. We would ask for a new road into our facility, widening north-south roads, fixing the intersection at New Haven and U.S. #1, and making the Melbourne Causeway into four lanes with a high rise bridge at the center. To put teeth in our plan, I hired a real estate agent, and swore him to secrecy. His job was to take options on about 250 acres of land abutting Radiation property. The largest piece, a quarter section or 160 acres, was square and directly across Palm Bay Road. Altogether we paid about $400,000 for the options. The options had a time limit of 90 days for exercise. We wanted it that way to put pressure on decision-makers when we made our actions public. Then I made my first (and only) entry into the political arena; not as a candidate, but as an individual trying to influence government. I sent Robey Green in his flying machine (Aero-Commander) with a photographer, to fly over the congested traffic at rush hour. Their instructions were to get pictures of many roads congested by traffic during our rush hour. To make sure we had what we wanted, Vergason went along and supervised the picture taking. We invited our Tallahassee legislative representatives, the Palm Bay City Council, the Melbourne City Council, Brevard County Commissioners, and key people from the Florida Department of Highways to a dinner at the Palm Bay Country Club. Attendance was good. We assured the invitees that we would give them information of importance to the community.

After dinner, I made a presentation to which we invited the attendance of radio and newspaper reporters. To remind the politicians what Radiation meant to the community, my slide presentation included Radiation employment and payroll figures. Further, we reviewed the amount of local taxes Radiation had paid over the last few years. Using averages, we summarized local taxes paid by our employees. Finishing the introduction, I pointed to the rapid growth of the company. Next, I defined the problem presented by the poor roads leading to our facility. Slides showing rush hour traffic in aerial view were effective in proving our difficulty. I discussed actions we had taken by using staggered hours and preferred parking. I flashed on the screen tables showing our recommendations. Then came the bombshell.

Before showing our situation to the community and to the investing public, we purchased options on 250 acres of land. However, we would not exercise those options until we had a commitment from various governments that they would solve our traffic problem. We could not continue our growth in Brevard County—or Florida for that matter—thus increasing our difficulty and that of the community. If government did not improve our roads, Radiation would have to take its growth elsewhere. Time was important. Our options were good for 90 days. The end.

The presentation was not designed to produce applause, and it succeeded. There was dead silence when I finished. At meeting's end we distributed a news release that carried the logo "IMPACT!" with three company names, General Development, Soroban, and Radiation. In part the release said that our three companies planned to " ... triple their economic impact on South Brevard by 1973, if the environment is right.... The future is up to you and others, Harold O'Kelley told a gathering that included members of the Brevard County legislative delegation, state officials, county commissioners, public officials and area business leaders. "To enable us to grow as we should, the communities of South Brevard must remain abreast of our growth by developing the needed facilities and services. What we need now–and not later–is action. "O'Kelley outlined what [the three companies are planning in the way of growth between now [1968] and 1973. He said that during that time, the number of employees of the three companies could grow from 4,400 today to more than 11,000 in 1973, while the corresponding increase in payroll would swell from $33 million to $93 million annually, an increase of 181%...O'Kelley highlighted a number of goals he termed "necessary for the optimum growth" of the three firms. These included better roads and bridges, area-wide planning and zoning and unification of all or parts of South Brevard. Referring to a recent employee traffic study that showed an automobile count of 3,000 cars daily traveling to and from the Port Malabar Industrial Park, O'Kelley said top priority should be given to better roads and bridges in the area. With the expected rise in employment, the three companies could be putting more than 8,000 cars on the roads of South Brevard in 1973--creating an almost insoluble traffic jam." Attached to the news release was a tabular summary giving "Selected Community Impact Based on Current and Projected Personnel Growth." We loaded the attachment with numbers such as headcount (4,400 in 1968 growing to 11,000 in 1973) and Employee Income ($32.6 million in 1968 to $92.7 million five years later).

For me, the next month was a whirlwind of activity. Local politicians requested meeting after meeting and brought the Florida Department of Transportation people with them. They tried to determine the minimum that they could do and still satisfy our requirements. This negotiation went on and on and on. Newspapers ran stories supporting us. The editorial pages took up our cause. Radio stations did their part. To further increase the pressure on local politicians, I visited the governor. While there, I made a presentation that was a condensed version of what I said at the country club. Governor Kirk said that he would support us in every way that he could. The result was that the Department of Transportation committed to building a road from our facility, going north. It would include a high-rise bridge over the railroad. Moreover, it would intersect US #1 south of Melbourne. Further, the Department moved forward their plans for building a four-lane high rise bridge at the Melbourne Causeway. Brevard County committed to four-laning Babcock Avenue and Palm Bay Road. And Melbourne and Brevard County officials committed to solving the congestion at the intersection of New Haven and U.S. #1. Radiation (Harris) then bought 160 acres of the 250 under option. (The 160 acres was a quarter section. Thus, it was square.) This land is on the North side of Palm Bay road. Harris Semiconductor buildings now stand on the property.

Of the three companies, only Radiation (Harris) achieved the forecast results. Soroban later closed and I led an effort to purchase the building for Harris. General Development has declared bankruptcy. In any case, the county desperately needed the infrastructure and so did Radiation.

Ernie Ploeger Remembers

Ernie Ploeger cites the reason he was hired at Radiation (in the accounting department) as follows:
He grew up on the farm in Michigan, and one of his responsibilities was managing the stock of beans. So when Homer interviewed him and learned of his background, he was immediately made an offer because Radiation needed a bean counter.

Frank Perkins Remembers

My first job after getting out of the army in 1954 (and my first an an electronic engineer) was at Melpar, Inc., in northern Virginia. Melpar has a strong connection with Radiation, as the employer of Homer and George when they founded Radiation. Melpar was a strong technical company, but the management was peculiar and not sensitive to their employees, and the company had a high turnover, including many who became Radiation employees. I knew Scott Campbell and Bill Vernon there. After about a year, I moved to the warmer climate of central Florida.

My first Florida job was with RCA Service Company, who did the engineering for the Air Force Missile Test Center. Brevard was growing rapidly in the mid fifties, and housing was hard to come by. I finally found a rental house in "Old Loveridge" in Eau Gallie, a few blocks from Ballard Park. Traffic was terrible, and everyone car-pooled where possible. My five-man pool had three people named Frank, including Frank Lewis and Frank O'Kelley, Harold's brother. The route led up two-laned A1A from Canova Beach to Patrick AFB. At the time, there was no development along this road points—only endless palmettos.

It was nice being in Florida, but the RCA job was boring, and I eventually got a job at a small electronics company in Melbourne, Missileonics, Inc., founded and run by Jim Coapman, an ex Radiation engineer. Missileonics was located in a big old wooden hanger at the Melbourne Airport. It dated from the WWII air station, and was located just east of where the terminal is now located. The work was interesting at Missileonics—we built Telemetry ELSSE tracking systems, and Interference Analysis Vans for AFMTC—but we never reached critical mass. As we were closing down, I finally got a job at Radiation.

Radicon/Radiplex

I was hired at Radiation (on November 9, 1959) to shepherd a family of standard products, including the Radicon analog-to-digital converter, and the Radiplex analog multiplexor. Almost every system Radiation built included these components and someone had decided it would be a good idea to utilize standard products in these systems. The units received a full product design treatment from an outside consultant, and were physically quite striking, with curved, gold anodized front panels. Electrically, the units had been breadboarded, but had never been tested as complete units, particularly to their full specifications. The designers were all busy working on new systems and were not available to support the standard products. It was my job to turn this raw work in progress into a useful product. To further complicate the issue, the "standard" Radicon was designed to operate in many modes and configurations, and was therefore very complicated (and expensive).

The units had been designed into a number of systems under development, but all of the project engineers had insisted on simpler configurations customized to their particular requirements. Furthermore, new semiconductor components were becoming available which allowed much more efficient implementation of some functions such as analog switches. The original Radicon used matched quads of diodes for analog switches. These required large reference voltages to make the switches sufficiently accurate. The Radicon regulated a 200 volt supply down to 150 volts for this purpose. The resulting power dissipation caused the regulator card to run so hot that you could burn your hand on the card edge. Later versions used transistor switches, in turn allowing a 8 volt reference with much lower power dissipation.

Eventually I managed to get the advances incorporated into a simplified version of the "standard" Radicon, and to make other changes, including use of a tunnel diode error amp to get this unit performing to full 12 bit resolution, but this was not easy. One interesting aspects of the Radicon program was that I had a good excuse to talk to with all of the project engineers using the Radicon in their systems. This provided exposure to the full array of interesting systems Radiation was building.

One of these systems was the "Pink Kitchen", for General Electric. This was a set of what we would now call computer peripherals, such as format converters, tape drives, high-s p e e d printer/plotters, A/D and D/A converters, etc. These were designed, like the computer mainframes of the day, to be displayed in a glass-windowed room to impress visitors, hence the stylish design and striking color scheme.

Another interesting system, built for White Sands Missile range, used a "hybrid" version of the Radicon to allow automatic scaling of multiplexed analog signals as they were digitized. This was desirable because multiplications were so time consuming in digital computers of the day. (The Radicon accomplished this multiplication by multiplexing the reference voltage in sync with the signal multiplexing, thus scaling the signal.)

Another notable system was for Norair (ex Northrup Aircraft Company). This system was used in flight testing the Norair F-5 fighter, which subsequently evolved into the T-38, used to this day (50 years later) by the shuttle astronauts. Working on this system was Bernard France, who became a close friend and lunch companion for many years until he moved to Washington state.

The largest single user of Radicons were the data acquisition systems for Thiokol Corporation, in Utah, for development of the solid rocket motors for the Minuteman missile. These systems were also heavy users of the Radiplex 89, the first low level multiplexor, which eliminated a number of complex amplifiers. The absence of multiple precision amplifiers allowed for a more compact data acquisition system, and the idea of a portable cart-mounted data acquisition system attracted the attention of a California subsidiary, Radiation at Stanford. It seemed simple enough, but the development was plagued with problems and delays, caused by inexperience in designing such systems, confusion in the specifications of the hardware and the immaturity of the hardware. I remember a number of emergency trips I made to the Pablo Alto area to solve problems in applying the Radicon and Radiplex. In those days the only jet service to California was from Chicago, so I was always faced with a plane change in the Chicago winter, before the days of jet ways. With balmy weather at both ends, I hated to wear an overcoat just for the transfer, but it was really cold on the outdoor run to the airplane in Chicago. After a few of these emergency weekend trips, I discovered that I could upgrade to first class in Chicago with my company Air Travel Card and felt fully justified in so doing. No one at Radiation ever questioned this expenditure. Eventually the California division ran out of money for this project and lost interest, and the unfinished system was brought to Melbourne and reassigned to the Products Division.

Tape Format Converter

As all this was going on I was assigned to work on the Boeing Tape Format Converter (TFC), intended for use in their development of the Minuteman ICBM. The TFC had an Apex serial tape recorder, which played back the recorded PCM telemetry bit stream for recording on computer compatible parallel tape on a pair of Apex FR400's. The pair of FR400's was necessary to allow continuous operation of the serial tape recorder. My job was the design of the control panel, largely devoted to managing the digital tape recorders. This required individually mounting and prepping the tapes, handling the automatic or manual switching between them, and automatically terminating the tape to suit the specifications of the computer.

I always regarded the TFC job, under the leadership of Larry Klingler, as an example of how electronic design should be done. Larry was a very smart individual, fun to work with, but with strong ideas on how things should be done. One of his pet peeves was control panel design. He strongly felt that system operation should be intuitive and that the design should minimize the chance of an operator error. I can still feel Larry roll over in his grave at the poor design of Personal Computer controls; he would never tolerate clicking the "Start" button to stop the computer.

His methodology as a project engineer was well thought out. He verbally described to me the functions to be performed with the control panel and then cut me loose. He regularly, but informally reviewed the design with me, having me describe what the operator did, and how the system responded. When the system was finished I felt very flattered when he complimented me on the design and operation of my part of it.

Subsequent formal efforts at design reviews degenerated into farces. A detailed written report would be generated and circulated to a panel of "experts". Then there would be a big meeting with view-graphs and handouts. This was all a big drag on the time and effort of the designer and by the time it all took place it was dreadfully late to make any design changes. I remember many times making a suggestion in a design review meeting and being told, "that's a good idea, but it is too late implement it." That made it seem futile to have the big presentation if it were too late to take advantage of suggestions. Truthfully, the objective was to check off the box that a design review had been held. In contrast, Larry's informal reviews made sure that his experience was incorporated in the design, but did not require a lot of time-wasting documentation.

Huntsville Sync Study

One significant event was my assignment to a PCM synchronization study that Radiation won from NASA-Huntsville. I don't remember being involved in the proposal or marketing for this program, nor exactly why I was assigned to it, but it had a notable impact on my career. For one thing, I became somewhat of the local guru in synchronization at Radiation. It also marked my first extensive collaboration with Dan McRae, an analytical genius in our Advanced Technology Department. This collaboration continued throughout my career at Radiation and Harris.

PCM synchronization was an immature science in those days. The NRZ format was the most bandwidth-efficient but had no discrete frequency component for bit synchronization. At least partially because of this concern, the Minuteman telemetry devoted 3 bits of 27 to a word sync pattern, for an 11% performance penalty. Someone had recognized that passing the NRZ format through a non-linearity would create a bit rate component for synchronization, but no one knew if this would provide robust enough synchronization. This was one of the questions to be answered by the Huntsville study. I breadboarded a number of synchronization approaches and showed that bit rate synchronization need not limit PCM performance. (It should be noted that the present day Internet suffers a 10% performance penalty through the use of a character oriented format, which can be viewed as a from of word synchronization.)

Frame synchronization was also an issue. People generally understood that the insertion of a periodic sync pattern in the data was a good approach to frame sync, but it was unclear what parameters of pattern length and frame rate were appropriate, and what exact sync algorithms would yield satisfactory performance. Dan devised analytical procedures that allowed analysis of frame sync performance in terms of format parameters and synchronization strategy and settings. The insights into bit and frame synchronization served Radiation well in winning a number of PCM Decom system jobs, which were a significant business area for us at this point in time.

Bit and Frame Synchronizers

I designed and built several bit and frame (group) synchronizers for various decom jobs. The bit synchronizer was particularly important because it affected the overall performance of the system. The bit synchronizer (or Signal Conditioner) accepted a noisy analog version of the PCM bit stream (from a radio receiver, for example) and output a binary (digital) bit stream and an associated clock signal. They were challenging devices because they had to operate over a very wide bit rate range (typically one bps to one million bps) and needed to synchronize and make bit decisions in a near optimum manner in the face of noise, jitter and other disturbances. I built a few synchronizers of PC cards (see "Cards vs Modules" in the Appendix), but modules offered the chance to build a compact modular synchronizer that was dubbed the Model 5220. My first version had a design flaw in the form of leaky selection switches for the loop filter, which caused flaky behavior under some conditions. These synchronizers happened to be for NASA-Goddard, who were often a difficult customer and this time happened to have a particularly antagonistic engineer on the job. He found the main problem in the units, but proceeded to nit-pick with a bunch of other mostly made up problems. He and I grated so badly that Radiation management wisely kept us from talking while I fixed the real problems. Eventually this NASA engineer came to Radiation to present his case for junking the 5220. I wasn't allowed to attend the meeting, but Dan McRae and others demolished his technical arguments and the updated 5220 went on to wide and successful service.

PCM Decoms

A PCM Decom accepts a multiuplexed serial stream and routes each channel to a particular output channel or device, such as a D/A converter. PCM formats were flexible and could be quite complex (see Appendix ""PCM Telemetry"). (The Gemini format even changed in flight.) Decoms need to be flexible to handle the variety of formats. Early decoms, such as one we built for McDonnell Corp. for the Gemini program, achieved this flexibility with patch panels, but the programming of these with patch cords was a complicated nightmare. A better approach was to have electronic storage of all the parameters and have the decom look them up as it need them. This was easier said than done with the components available in the 1960's, considering that the short time available for processing a word. A system built for North American Aviation used a rudimentary stored program decom, but the first system to really exploit the benefits of a stored program approach was the 540 Decom designed for NASA. It was used for testing and checking out the various systems used in the Apollo moon landing program. The procuring agency was the Manned Spaceflight Center at Houston. The marketing of the systems involved lots of maneuvering by enthusiastic marketing types, hiring away competitors engineering talent, but the result was a sophisticated system that performed its tasks well and was widely used at a number of plants and sites.

Bernard France devised an efficient technique for performing the necessary processing. On the early versions of the 540, I mostly remember the long hours of checkout, trying to get the stored program technology working, under severe cost and schedule pressure. In spite of the pressure it was fun because of the other engineers working on it, and the relaxed but effective management of Al Medler. Al would come in to be with us on the long night shifts and was comforting support even when he dozed off on the stool. We finally met all the specs and were rewarded with what to us a "production" contact for 20 or so systems, installed at Cape Canaveral, Houston (MSFC), Daytona Beach (GE), and Downey, California (NAA plant). In the later phases of the program I was the senior technical person working on the job and got to adjust a few details to suit my preferences. I wasn't the titular system engineer, but I got to do most of the fun management under Medler's gentle guidance. One fun aspect was that we were an autonomous organization, with our own leased building. We also had a great staff of engineers. I particularly remember Bob Whitlow, Gerry Duggan, Jim Pettigrew and Jack Wark. We built a number of other stored program decoms. The PCM-DHE, for NASA's world-wide tracking network, was built around the same time as the 540's, using the same technology. Later we built the SGLS (Space-Ground Link System) decoms for the Air Force.

In 1967, things started to change with the acquisition of Radiation by Harris-Intertype of Cleveland, Ohio, later to become Harris Corporation.

The purchase of Radiation by Harris-Intertype marked a transition from an Engineer-oriented company to a Manager-oriented company. Previously, almost all the managers came from an engineering background and engineers and their technology were the driving influence. Later, the philosophy developed that a good manager could manage anything, whether he knew anything about it or not. Whatever the business effect, this made the company a less fun place to work.

Remembering Peter Petroff
By Don Sorchych | Sonaran News/August 6, 2008

While at my first professional job after finishing college, with Radiation, Inc., I met many characters. Peter Petroff is high on the list.

Peter literally sailed, well motored, into the town of Melbourne, Florida on the intercoastal waterway

(called the Indian River in those parts) in a catamaran he designed and had built while working for an international construction firm in Vietnam.
He got there in a circuitous way.

Peter was Bulgarian with a pot belly, long jet black hair, penetrating brown eyes, a swarthy complexion and a heavy accent. He was the most pragmatic man I ever met and among the most amoral.

The obituary, which was published upon his death, differs in so many ways from what he told me and what I heard from mutual friends that I can only go by my recollections.
When he died, his old boss at Radiation, Jack Pruitt, called me and urged me to write Peter story as he laughed about the obit claim he had invented the electronic watch.

Peter told me he had been captured by the German army and conscripted into military service as a Nazi soldier. At war end, the Bulgarian government offered a short period of amnesty, but after that anyone who had served in the German army would be declared a war criminal.

The Canadian government offered a visa and he immigrated there. He joined an overseas construction company and while in Vietnam designed and built his catamaran. He had the catamaran shipped to Canada, and went south on the intercoastal waterway with his wife and two sons.

Peter was always lucky and after berthing in the Melbourne city harbor, his unique catamaran was noticed by George Shaw, the co-founder of Radiation.

George always had yachts and after a tour of the boat with Peter he pulled strings to get him a position at Radiation. This chance meeting gave Peter access to executive row at the company and he never ceased to use it.

I met him when I was leading a group that designed the telemetry for the Apollo project and several missile systems. Peter worked in the mechanical design group and concentrated on miniaturizing the size and weight of electronic modules. He may have invented a foam encapsulation method, which became a standard for our designs for a number of years. But he may have copied from somewhere too.

As a friendship developed he spoke of his days as a Nazi soldier with great pride. He thought Adolph Hitler was a legend whose only mistake was to show, rather than hide, his prejudice for Jews and people of color. He thought that it was "stupid" of him but not wrong.
He told me he figured out the American system. He said, "Donalt, all I haf to do is mofe jobs enough times und get 15 percent each time und I can double my vages in six mofes."

He soon left and joined a Radiation spin-off in Orlando, Florida with a 25 percent boost in pay. He took with him bundles of designs and know-how. Although there was a furious exchange of letters from attorneys of both companies, nothing ever came of it.

After a year or so I received a call from a NASA recruiter. Did I know Dr. Petroff?" I replied, "I know Peter Petroff." She wanted to know more about the research projects he headed. I explained he designed mechanical modules for electronics. She was adamant that he managed major electronics research projects and was quite upset I didn't agree.

Next day a distressed Peter called me and asked why I didn't lie. Having no answer from me he asserted, "You are a goddamned Boy Scout!" and hung up.

Nonetheless, he was appointed to Director of Research at NASA in Huntsville, Alabama.
While there he negotiated contracts with companies, ostensibly to monitor vital functions of astronauts.

Later he founded Care Inc, a patient monitoring company using the technology lifted from the NASA contracts.

He and his lovely German wife Helen came to visit after incorporation of Care Inc. in a brand new Lincoln. He proudly showed me his list of stock holders and his majority holdings, including some of Werner von Braun's scientists.

I am sure Care Inc. went belly up, but his obit said it "became" Electro Data and it was there that the first electronic watch was invented. I don't think so, and further, if it was, Peter had no electronic skills. None.

Later, he and his sons started an environmental monitoring company. That was the last I heard from him.

His obit states: "He joined the space projects carried out by the precursor of Harris Corp, (Radiation, Inc.). He helped design weather and communication satellite systems and organized the company's semiconductor division."

Not exactly.

Peter designed mechanical assemblies. Period.

And as far as the semiconductor division, which I later managed, it never would have existed without Peter. Peter was viewed as an expert in microminiaturization. And he correctly argued integrated circuits, then in their infancy, would be the ultimate solution in future systems.

With his connections in the executive suite, he badgered and bullied them, virtually wearing out the carpet in visits followed by lunches in the company cafeteria. He was the ultimate salesman, bright, persuasive and persistent.

His heavy accent helped too. His friendship with company co-founder George Shaw was key as George was a sizable share holder and equally persistent. Executive row rocked with dissent, but Shaw's booming voice could be heard arguing the strategic necessity.
It was a huge risk for a small company, which at that time was a 100 percent U.S. government contractor.

Finally the decision was made to proceed, solely due to Peter's persistence. So, although he never "organized" the division as stated in his obit, it wouldn't have happened without him.
I have already written the next chapter, the hiring of Uryon S. Davidsohn to head the semiconductor division and his ultimate demise and ghastly death.

Peter, I am sure you conned St. Peter to let you through the pearly gates to the heavenly realms.

R.I.P.

Don's editorial triggered the following letter: from Bob Hoss of Cave Creek, AZ.:

I have some interesting information for you that might complete your article on Petroff and will fill in the real story regarding Electro Data and the world's first electronic watch and Petroff's role in all that.

Your article caught my attention because I began my career at Cape Kennedy and knew many of the folks at Radiation and Harris and indeed Petroff crossed my path as well. After the space program I became one of a small number of engineers who made up the company called Electro Data in Richardson, Texas.

In 1970 Electro Data did build the world's first digital watch, the Pulsar – for Hamilton Watch company

– BUT the development had nothing to do with Petroff who didn't enter the picture till after 1971. It was developed by the then President of Electro Data and an engineering manager from Technology Incorporated (unfortunately I can't recall their names).

After reading your story it may be no surprise to you that Electro Data was doing quite well until we were forced to merge with Care Inc. – which took us both out of existence. The story in the press was that Care was in financial trouble, so bad that it had already been refinanced several times and had built up "such a tremendous amount of debt that it was unattractive to its investor." So their board, which had some common investment interests with Electro Data, forced a merger in 1971 (we understood that it was to use our good cash position to shore up Care). But in doing so they voted the president of ElectroData out (the one who had championed the digital watch effort) and put president of Care (Petroff I assume) in as chairman over both companies!!

Indeed he must have been quite the salesman! A year later they reversed this and sent him back to president of Care, but by then the damage had been done – we were both now in the red. So no, Petroff had nothing to do with developing the watch which was done a year before he took control, and the only contribution of that merger seemed to be to destroy the company that had achieved that marvelous goal.

Side note: Interestingly the Pulsar site credits a guy named Wuischpard with designing the Pulsar – but he only designed the case! The true invention of the watch itself was by two unsung geniuses at Electro Data (one who was pushed out when Petroff took over) – two names unfortunately I can't even recall after all these years.

Joe Pira Remembers

Joe spent four years in the Air Force, including a year in Korea. He graduated from the University of Florida with a BME and came to work for Radiation in August of 1959.

His first boss was John Mazur.

Joe worked on numerous programs, including:

Ground Pilot trainer for the Bullpup missile.
The Bullpup was an early air to ground missile, which was guided by the pilot visually tracking a flare on the missile and transmitting corrective guidance commands on an RF link.

The trainer consisted of a 17" CRT which displayed a simulated missile flare and accepted guidance commands from the trainee. The Trainer had severe environmental requirements, including shock, vibration, low and high temperature, pressure and humidity. It was designed and built in Orlando for the Martin Company, contractor for the missile system.

OAO.
The Orbiting Astronomical Observatory (OAO) satellites were a series of four space observatories launched by NASA between 1966 and 1972, which provided the first high-quality observations of many objects in ultraviolet light. The success of OAO increased awareness within the astronomical community of the benefits of space-based observations, and led to the instigation of the Hubble Space Telescope. Radiation built the spacecraft data handling equipment for Grumman Aircraft Co. The equipment accepted both analog and digital data signals from various equipment in the satellite and processed them into the proper format for storage and transmission to a ground station.

ASW-25

The AN/ASW-25A Communication Set, Digital Data System was an essential link in the U.S. Navy's All-Weather Carrier Landing System (ACLS). This system provided a highly reliable capability for safely landing carrier-based aircraft despite adverse weather, sea, and visibility conditions. The Final Report was dated December, 1969.

Apollo

An interesting note: The specification provided to Radiation did not allow sufficient space for the specified channel data. At 3:00 AM in California, Joe noted a 1 inch error in the specification versus the actual hardware. The result was more space and full data channels achieved.

Concorde

Radiation built electronic instrumentation for test versions of the Concorde Supersonic airliner. The first four aircraft were purely test vehicles and well described as flying laboratories. Each of the prototypes carried about 12 tons of electronic test instrumentation, much of it specially developed for the purpose. This instrumentation was capable of recording measurements of 3,000 different parameters, including pressures, temperatures, accelerations and attitudes, and the information was recorded on magnetic tape in the aircraft for later analysis in the ground data processing centers. While the aircraft was in flight, certain basic information was continuously telemetered to ground monitoring stations.

1300
The Hexagon photo reconnaissance satellite

Xerox Copier design and production.

Study for an offshore manufacturing plant. This work involved trips to Mexico, Haiti, Honduras and Brazil.

Manager of Programs at the Miami Lakes Operation.
Transferred to Miami Lakes facility as Manager of Programs and later as Director of Programs until a decision was made to close the Miami Lakes plant.

Jack Pruitt's Memories

It was early spring in 1956 and the corporate recruiters descended on Georgia Tech. DuPont set the early standard for job offers with a starting salary of $400 per month for BS engineers--a little later the aircraft companies upped the ante to $450 to get people to move to California. Jack started making the rounds of the company booths, and took a liking to Western Electric (AT&T) and Allis Chalmers. He visited both and became set on AC, even though it was in Milwaukee. Pat gave him a big "raised-eyebrow" with that announcement, so Jack gathered that he should keep looking. Luckily, he didn't have to look too far, as he had an invite to visit a small electronics company by the name of Radiation, Inc. in Melbourne, Florida. They flew him and one other Tech engineer to Melbourne in their corporate plane. In addition to the interview and plant tours, they took him to lunch at the Bahama Beach Club on the ocean, where they had arranged a seaside table. That about did it. He took the job that was offered, with a salary of $415 per month and a reporting date of July 9. He would be working for Wally Silver, the Master Mechanic (meaning head of the machine shop, drafting and chemical processes). That turned out to be a good way to get a feel for industry, and it prepared him for Design Engineering, where he would land after a 6-months stint of active duty with the National Guard.

Jack's first boss was Wally Silver, a quiet but firm man who was responsible for drafting, chemical processes and the machine shop. The supervisors of those three departments became Jack's mentors, teaching him the right way to do things and ways to get things done. They were experienced, dedicated and willing coaches.

Jack had played basketball in high school and intramurals in college. He attempted to keep it up once he got to Melbourne, but the rigors of practice while trying to get a new life at work and at home got to be too much, so he "retired."

It was post-Korea times but the country was keeping its military strength ready for whatever might come next. Jack chose to join the National Guard. The unit was located in Cocoa, and held weekly evening meetings. There were also monthly weekend bivouacs and two weeks of training each summer. His enlistment lasted about a year, before he was called to active duty and assigned to Fort Jackson, South Carolina.

"What's that ring on your finger, soldier?" the Inspector General demanded. "It's a college ring, Sir."
"From where?" "Georgia Tech, Sir."
"Uh huh. That's a pretty good school. What'd you study?" "Mechanical Engineering, Sir."
"You graduate?" "Yes, Sir."
"Mechanical Engineering, huh? What do you do for us?" "Truck Driver, First Class, Sir."
"Damn Army!" the General said under his breath, walking away, shaking his head.

With a lot of waiting time in the field while the troops were doing their thing, Jack decided to take the Series 10 Course to become an officer. Completing the academic courses, along with his active duty service would qualify him for a commission and a second lieutenant rank. He completed the course and was awarded a certificate. But, with peacetime in full bloom, the Army found itself with too many officers, and cancelled the Series 10 program. That was one time when "seizing the day" didn't work. Jack remained a non-com for the balance of his National Guard career.

As Jack returned to Radiation from active duty, he found conditions had changed. Wally Silver's budget had been cut and Jack's old job was no longer available, but, once again, something better showed up. He was assigned to the Mechanical Design Section—a move from manufacturing to engineering—and a progressive step. His assignments included designs for packaging a Doppler radar system for helicopters, and servo controls for large ground-based antennas. These jobs lasted about a year, and as it would turn out, were the only assignments in Jack's career as an *individual contributor*, in other words *true design engineering!* He would quickly find his natural fit in supervision and management; and concluded that he was, in his words, "not a very good engineer."

As the radar and antenna projects wound down, Jack was assigned to the Product Design Section, headed by John Walker (a former fraternity brother at Tech). John's section's major project was the PCM telemetry system for the Minuteman missile, under contract with Boeing Aircraft. Product Design was responsible for packaging the electronics to withstand the severe environments of missile launch and space. The project was well along when Jack was assigned to take charge of the Product Design task. Bill Corry, a talented mechanical engineer, had conceived and led the design from its inception. Jack's job would be to shepherd the project through environmental testing and into manufacturing for the production phase, neither of which would turn out to be easy.

The Minuteman project was a major milestone for Radiation. The technical specs were stringent, the customer (Boeing) was tough, and Radiation had never faced rigid quality control and production requirements. The teams worked around the clock, including Saturdays and Sundays. But Friday nights were a time to recharge, and the gang worked even harder at that—meaning getting falling-down-drunk, at times. The parties were held at various homes and, as they grew increasing rowdy, the wives declared their homes off-limits, pushed the parties outside, and allowed only one-at-a-time entry for the bathroom.

Jack's boss, John Walker, resorted to bringing an oxygen tank into the office on Saturday mornings to revive his people.

The Melbourne area was like a newly discovered oasis to the Radiation people—and to the people at Patrick AFB and The Cape. The military, NASA and Radiation were hiring dozens of people weekly, and they were all "from somewhere else." As a result, there was no family to fall back on—or to interfere. Nobody went home to mama on weekends—it was too far. And, nobody judged you by your family's reputation. It was, "up to you, Charlie," so to speak. This meant freedom and responsibility that was absent the family ties that had been so much a part of their prior lives. It meant that what you accomplished--not what your daddy did--determined where you "fit" in the new structure—both at work and in the community. At work, the men competed for respect and promotion; in the community, the wives got active—both to retain their sanity and to make friends. Volunteer organizations prospered. The hospital, with its Pink Ladies, and the Junior League, became both vibrant civic benefactors and the core of society.

Radiation slew the dragon. The company succeeded in getting the Minuteman system through the rigid specs and into production. That success put Radiation in the big leagues. It became the premier aerospace telemetry company in the world. Just as the company was turning the corner on Minuteman and Jack was gaining success as a project manager, his boss resigned. The boss, John Walker, broke the news to him, and added that his boss wanted to see him. As it turned out, John and his boss Dick Hultberg had already discussed Jack being promoted to take Walker's position. It was a very important slot, and would be a challenge for Jack. As shocked as he was, he nevertheless realized that it was exactly what he wanted. So, he accepted and became Section Head of Product Design, Aerospace Division.

There was little time to celebrate. New opportunities for the company were occurring everyday it seemed. The word about Radiation's accomplishments had quickly spread within NASA and the Air Force. One of the first bid requests that the company received was from NASA for the Nimbus satellite. It was followed almost immediately by a request from Bell Laboratories for their company-financed Telstar satellite. But the specs were too severe for the current packaging design. A new concept would be needed.

The answer arrived—from Bulgaria! He was an immigrant by the name of Peter Petroff. His tenure would be short, but he would sting like a bee, with the venom of a viper—for the competition, that is. Somehow, Petroff had showed up at the company's front door and managed to talk his way into an interview with Vice-President George Shaw. Shaw saw immediately Petroff's potential, but had no idea where to put him. Not being a citizen, he could not work on any classified jobs, and he was not an electronics design engineer—he was more of a physicist and instrumentation expert. So, he decided to "give" him to Pruitt, who was told to "see what you can do with this man."

Jack had no idea what to do with him, but he showed him everything that was going on, gave him a bunch of "packaging" magazines to read, and then the specs for the new proposals that were just getting underway. Petroff read everything he could get his hands on and then managed to find a systems engineer whom he bombarded with questions. His questions got the engineer's attention and pretty soon he took Petroff to meet top circuit designers, Bill Eddins and Hank Patterson. It was there that he learned the deeper technical principles that would drive many of the constraints that the packaging guys would have to meet.

Immediately, Petroff asked Jack for a technician, one who could turn ideas into hardware. He was told he could "borrow" Bill Stankos, who was already working full-time on active projects. It was a match from heaven. Bill knew in just a few minutes that this man was different. Pete laid out for Bill the three major design concepts that would be required, if we were to meet the specifications of the new proposals. They were:
· Welded cordwood modules
· Foam-in-place encapsulation
· High-density printed circuit boards.

Stankos had never worked with any of those, but he quickly grasped their significance. He advised Pete that Product Design could handle the first two, but that they would need Manufacturing to take the lead with the circuit boards, and suggested a meeting with Phil Derrough, manager of Special Processes. Within days, welders and a range of foam-in-place encapsulants were on order and Phil Derrough had been brought up to date on Engineering's vision for the next generation packaging design. Phil beamed as if he had won the lottery. He had been "boot-legging" the development of multilayer circuit boards for 2 years, and now his vision was coming true—and, he would have a customer to pay for "legitimate" development. Next, through intense lobbying, Pete was able to get Jack to commit Stankos to him full-time—which meant that Stankos was now committed full-time, twice! Pete then coerced the Drafting Manager to provide a designer to produce sketches for mechanical breadboards, and he returned to Circuit Design and pleaded for schematics and parts lists so he could begin to size the packages and the interconnections. Within 2 weeks, Pete had a loosely connected cadre of people from a half-dozen departments working with him--while still carrying on their assigned duties.

Pretty soon the welding machine showed up for the assembler that Stankos had commandeered from Manufacturing, along with a large box of faulty components. The next day she was interconnecting components by welding nickel wire to leads, and Bill then put them into a cordwood shape to show the concept. He had a friend in the machine shop hog out a dozen aluminum cavities that he would use to encapsulate the welded module. Two days later, Petroff walked down to Don Sorchych's office, laid a welded module on his desk, and leaned back in his chair. Sorchych, a brilliant, aggressive digital design engineer was slow to show approval, and immediately began tearing into the concept. But, Pete had already sized up Sorchych and knew both the questions and the answers. In short he said, "Product Design can design and produce these, but we can't succeed without you." You've got to teach us what it takes to make them work flawlessly. That was music to Sorchych's ears.

One thing he demanded was subservience from all other functions, including Product Design. Now that Peter had pledged that, Sorchych had no defenses. With Sorchych's backing, Peter would have credibility, but more importantly, he would have Electronics Design's active involvement. The barriers came down. Departmental jealousy evaporated. The Competition was now in trouble!

From there, the opportunities grew. The company won the contracts for the Bell Laboratories Telstar and NASA's Nimbus satellite. The new packaging concepts had worked. And, a fourth dimension had been added. The engineers had realized that to meet severe vibration environments the package could be designed as a "solid mass." This was made possible by designing all modules to be of the same height; stacking the module assemblies on top of one another, like layers of a cake, and putting the entire stack under compression inside a very lightweight box. It worked beautifully. Nothing was left "flapping in the breeze" to shake loose. The Competition was, indeed, in trouble.

Winning the Nimbus contract had required an additional piece of ingenuity, however. Very early, it was realized that the unusually large number of incoming wires to the package would exceed the capacity of the circuit boards. The main problem was getting 1000 input/outputs through a bank of connectors on 10 printed circuit cards. After several blind alleys, Pete rushed into the office one morning literally beaming. "Jack, Jack," he said, "I've got it. Let me show you!" With that, he pulled some toy blocks from his pocket and told Jack that he found the solution when he saw his children playing with blocks. We could double the number of PC cards and double the number of connector pins simply by attaching half the modules to each card of a pair and then nesting them face-to-face checkerboard style. It did the trick! The project became a big success. Petroff again found a unique way to solve a seemingly impossible problem.

During the ensuing years, the company added major contracts for the Apollo capsule, LEM (Lunar Excursion Module), Titan missile, Lunar Orbiting Satellite, two major classified satellite programs and several mission-critical strategic satellite networks.

That engineering team, aided by the strong leadership of Phil Derrough in Manufacturing, went on to develop multi-layer printed circuit boards for greatly increased component densities—and thereby dominate the PCM telemetry market for space flight. Jack got magazine coverage during that era as an expert, but it was really the engineers and technicians who made the concept a success.

Peter Petroff moved on, relocating to Huntsville and going to work for NASA. Five years later, he and his three sons started a company to digitally measure and monitor water quality in fresh and refuse water systems. After that, they formed a company that pioneered Ultra Wideband transmission for radio and telephony. Jack and Pete remained friends, and Pete remained Jack's most important mentor—a feeling that he expressed years later at a company reunion—in Pete's presence.

The Manned Orbiting Laboratory was an Air Force program conceived to put military personnel into Space for intelligence and, perhaps, combat operations. Harris Corporation won contracts to provide sub-systems for the laboratory. Jack left his post as head of Product Design and was appointed Program Manager for those contracts. (This was another *carpe diem*, and a decision that he made against the advice of his confidants.) McDonnell Douglas, now Boeing, was the Harris customer.

The contracts, which were fixed-price, had been fiercely competed, and Harris's winning bids presented a profitability challenge to the program team. However, the customer almost immediately began making changes, which required extensive proposals, followed by negotiations. Jack began to run into difficulty with the customer's representative, Lew Hansen, and grew increasingly frustrated with his inability to convince Lew to accept his arguments and justification of price.

One day, as he was bemoaning his problem to his boss, Bill Vernon, Bill asked, "Is your written proposal self-explanatory, and is every number backed-up by rational justification?" "Well, I think so," Jack responded, "and where it isn't, I provide an explanation during our oral negotiations." "I don't know the man," replied Bill, "he might be lazy, incompetent, or smart like a fox, but he's manipulating you. Oral supplements are fine in medicine but they have no place in negotiations! Make the proposal stand on its own!"

Jack took this in, and reflected on previous negotiations. Maybe he was depending too much on verbal persuasion. Maybe he needed to redesign his negotiating technique! So, Jack embarked on a brand new proposal scheme. Each element would be self-explanatory; it would be supported by cost, schedule and performance explanations; all numbers would be further substantiated by experts in the organization; and, everything would be neatly tied together in easy-to- follow cause and effect descriptions. There would be no need for further oral explanations!

It worked like a charm. The next group of proposals was submitted and, this time, instead of Lew calling for a negotiating session, called to say they had been received and that we would be hearing from him. After another week or so, Lew called back and say that he saw no need for getting together, i f we would accept his offer on price. He would make an offer; Jack would counter offer; and, after a little more cat and mouse, they would settle. Lew went on to say that our proposals showed marked improvement, and that his specialists had found no problems with our justifications.

The lesson learned, especially with bureaucratic organizations, is that the guy across the table is often not the decision maker. Don't expect him to accept your arguments and then go try to sell them to his management. Do that job for him. Make the written material justify your position. Your adversary becomes a messenger, but a willing messenger. He'd much rather put his management on the spot to find holes in your arguments than to do that himself. And, besides, it's much less work.

It is 1969, and federal budgets are tight. Word comes down from the Defense Department to put the MOL program on an extended schedule--not a termination, only a delay. (But it was an omen of things to come.) Jack's team put together their estimates for supporting the program in a barely sustaining manner. Jack summarized the estimates and went to Pricing Committee for approval. This committee consisted of Bill Vernon and Bill Premaza. They listened to Jack's presentation, asked him to allow them to discuss it, and called him back in a little later. Their instructions were brief: "You do not have adequate effort in the proposal to support the program. We want you to triple the numbers." Jack sat in stunned silence. He had already been worried about justifying what the team had asked for—now he was being asked to triple it! Impossible! His immediate reaction was to challenge the pricing committee to 'justify" its logic. Premaza's answer was brief: "That's your job. You'll think of something."

For the next several days Jack brooded about the prospects of developing a sane proposal for the customer. He discussed it with several associates but every one just frowned and walked away. Then, on Sunday afternoon at the dining room table, it began to come together. He determined that each functional department involved in the program would maintain a small staff on the program, in order to maintain knowledge and continuity. He termed it the "marching army." And when he was finished tallying the numbers, low and behold, it matched the amount that the pricing committee had decreed!

But, that was not the last big exercise on MOL. Six months later, the program was killed. This resulted in another negotiating challenge for the team. The ASPR regulations provide criteria for paying a contractor whose contract is cancelled for convenience by the government. The first step is to ascertain if the contractor's program is profitable or not. If not, the government will only pay the contractor's costs, less the Percentage that his loss amounts to. If the program is profitable, even a miniscule amount, then the fee awarded on top full cost reimbursement is based on such things as the value of the technology contributions made on the contract.

The MOL team was able to show convincingly that the contract was profitable. Once that hurdle was overcome, the Company qualified to receive a substantial award fee due to the important technology contributions made. As a result of these accomplishments, Jack was promoted to Director, Aerospace Programs.

The place was abuzz one morning in 1967-. Word had leaked out during the previous evening that Radiation, Inc. was being bought.....by some Cleveland company by the name of Harris Interstripe—or something like that. It turned out to be Harris-Intertype, but then it was revealed that they made printing presses! Why on earth, the scuttlebutt had it, would we want to sell to somebody like that??!! Radiation had been in the news as an acquisition target, off and on for years, but mostly to aerospace or defense companies. Certainly not to a printing company! Well, it wasn't a printing company; instead, it was a pretty good sized machinery manufacturer. The word came down that the Company had signed an agreement to "merge," into Harris-Intertype, and that a more-detailed meeting would be held promptly to explain it.

Years later, Chairman & CEO Jack Hartley would explain to a group of Old-timers what had led up to the association with Harris-Intertype. He said that Radiation had become an ongoing acquisition target of several large companies, but that management was not too keen on any of them. For one thing, the Company would lose its independence and, secondly, it was not clear that any of the suitors could add much to Radiation's capabilities. However, management knew that Harris-Intertype knew nothing about Radiation's business and would probably leave the Radiation management alone. And, the upside was that Harris could provide financial strength to help with technical and facility investments that might be on the horizon with the rapid advances being made in semiconductor technology. But, also, Harris was well known for its management systems and its management development programs. So, Harris was considered the kind of merger that looked like it offered important plusses, without any minuses. They turned out to be right in their assessment.

So, with Radiation's record of success, why did Homer decide to sell? (He was the major shareholder, and it is not clear that his associate, George Shaw, concurred in the decision.) The answer is that he felt, having tasted the level of investment required to enter the semiconductor field, the Company would be stretched thin and would be required to take on extreme risks, financially, if they were to play a lead role in their markets. Was Homer right or wrong? With what he knew and with the prevailing opinions of the time, he was wise. However, the tenet that "systems companies" had to have their own semiconductor operations prove to be false. So, his premise that the company would need huge funding for semiconductor R&D and manufacturing was over-stated. Radiation did, in fact, invest heavily in semiconductors, but it evolved into a stand-alone business, and not an integral part of the "systems"

In the early years of the Semiconductor Division, "systems" projects were pushed to "design-in" circuits that were being developed by Semiconductor. It was not an easy sell for management, because the "systems" guys grew up making decisions that best suited their own requirements.

As the semiconductor companies evolved, they gravitated into specialties, such as digital logic arrays, analog circuits, amplifiers or power devices. So, "systems" engineers returned to doing what they do best, namely, evaluating what's available against their requirements and choosing the optimum for their project. Only rarely, did they have to resort to internally developed custom circuits to accomplish their goals.

So, did the merger into Harris-Intertype prove to be fortunate, or otherwise? It probably helped Harris more than Radiation, as a result of the management talent at Radiation that went on to control the combined corporation. Radiation CEO, Dr. Joe Boyd and Vice-president, Jack Hartley each went on to become Chairman of the Board/CEO of the Company. But, Harris did introduce valuable management concepts and practices that survive to this day. These included formal annual and strategic planning, management development/succession processes and management incentives systems.

Harris-Intertype brought an emphasis on management development. One example of that was the Senior Executive Development Program. That program provided a range of personal development seminars for the selected participants. One of these consisted of a retreat for a long weekend at the American Management Association's enclave in Syosset, New York. An element of that retreat was a "sound off at the boss," session, in which participants were asked to submit written anonymous comments to George Dively, the crusty Chairman of the Board. Upon opening and reading aloud one particular comment, Dively paused and then stammered, "Who the hell would say such a stupid thing?" The room could not hold back, and erupted in laughter. Without hesitating, Harlan Carothers, a well respected department manager, stood and replied, "I did sir!" thereby adding greatly to the admiration his associates already had for him.

The product and technology synergy that Harris Chairman Dively foresaw never evolved. The business units of Radiation grew in their established markets; and likewise, the Harris-Intertype units stuck to their traditional businesses. There was a modest amount of technology crossover from electronics to printing in the form of press set-up and ink control; and, the Harris Composition business transferred entirely into the electronics side of the house. But, both it and the entire printing equipment business were eventually sold.

George Shaw Remembers
(From Harris FYI, 1995)

George S. Shaw was co-founder of Radiation Incorporated in Melbourne, Florida, in 1950.

Many present-day Harris Corporation employees who worked with Shaw speak of him as a colorful personality and respect him highly as the foundation of the company's electrical engineering strength and inventive culture. It wasn't easy to attract some of the best engineers in the country away from established technological centers, so Shaw became actively involved in the development of Melbourne from an agricultural community of 5,000 into the Space Coast's technological and commercial hub.

Homer Denius and I decided to locate the company in Melbourne so we would be close to our customer at the Cape and have easy access to service our equipment. I suppose that the reason it was relatively easy to get the company started was that corporations were so busy recovering from the disorientation from World War II that they didn't want to engage in military work. The United States was involved in Korea, and they needed us.

The type of contracts we acquired just built one on the other. That speaks extremely well for Mr. Denius as a salesman. He was without a doubt a number-one salesman, and he had Bill Dodson, Jack Hartley, and others to help.

The biggest problem I had was getting people. We were engineer-limited in the beginning, and the early development of the company was dependent on our getting good people. The vehicles for doing that were the consultants who worked for us, and getting people to visit and find out what a lovely place Melbourne would be to live.

The consultants kept our people stimulated. Henry Zimmerman from MIT came frequently, and we would send our people up there to visit him. Dr. Larry Rauch and Myron Nichols, from the University of Michigan, were down here often. They wrote a book on digital telemetry, and a lot of it was from papers here.

Very early on, the need was seen for a school here, and so the Florida Institute of Technology was started by Dr. Jerome Keuper and of course Radiation supported it one hundred percent. I've been a board member constantly since FIT was started, and am probably the longest-serving voting board member. We have a lot of honorary board members but I still go to meetings and vote — and not always "yes."

Homer Denius and I both had families when we came down here and started the company, and they became very active in the social life of the community. We had the Radiation Wives Club very soon after the start of the company; they got involved in the League of Women Voters and the library. Many of the wives were active in the Sweet Adelines, and we started a Toastmasters club to help our salesmen and engineers in making presentations.

In those early days, we were sort of a company family. Six days a week we would work very long hours, and on Sundays we'd soak up the Florida sunshine and all get out on the Indian River with our families, sailing and swimming, or go to the beach.

We could see right away that some thing had to be done for the community in order to make it attractive to bright, young people coming out of MIT, or Georgia Tech, or Purdue. How in the world do you get them to move to a town of 5,000?

A first step was FIT. Another one was the hospital. The hospital was kind of like a 20-room motel down on Route One. I was on the hospital board with Jimmy Holmes, and the Holmes Regional Medical Center now exists because of our work and our pushing in that direction.

We built the community from the inside. I was commodore of the yacht club, to give people something to do and a place to go. It was very inexpensive then, and instead of the professional chefs and fancy meals of today the club members put on parties and cook-outs. There were boats, but they were little sailboats for families to have fun with.

Then there was the matter of the bank. If there had been a pawn shop it would have had the business. It was a pretty poor bank. So Mr. Denius and a group of local businessmen started the First National Bank of Melbourne. The bank was responsible for starting the Trinity Towers development here for the elderly, and it also helped finance FIT. The First National Bank of Melbourne was sold to the Sun Group after the merger of Radiation and Harris-Intertype, and the Sun Bank continues to be one of the principal banks in the area.

There was one crucial point of conflict with the community. We had acquired a contract to digitally instrument some aircraft to fly through nuclear tests — an extremely important contract for us. The planes could be trucked in, but after we did our work they had to be flown out intact. The runways at the Melbourne airport were too short. We offered to pay for extending the runways in exchange for use of some vacant World War II buildings we needed, but the city wanted us to pay rent, too. One of the difficulties was that it was impossible to talk about what we were doing. The project was highly classified. All I could say was that we have to have the runways longer. I couldn't tell them why. Fort Lauderdale, Orlando, and Miami were willing to do almost anything to get us to move there. We wound up building our own facility for he project in Orlando, where the runways were long enough, but we almost left the Melbourne area.

At that point, we could see that if we were to keep up with the business we were getting we had to get into some substantial and adequate facilities. Neither Mr. Denius nor I wanted to move to Orlando. The people in Palm Bay made us welcome and helped us find financing, and that's how we came to locate our main facilities there.

We ran a full-page ad every month in Fortune magazine. We talked about digital communications and how it would lead to satellite communications, personal wireless telephones, world wide instant communications, and extension of the American philosophy in the English language throughout the world, just as it's happening today. I guess we were thinking 30 years ahead, but that's what it takes. Some very good engineers saw those ads and came to work with us to help achieve those things.

By the time Dr. Joe Boyd joined us, we had plenty of PhD. 's in the company — maybe 30 — so the new Ph.D. in town was not the first. And we had lots of people with master's degrees and hundreds of graduate engineers.

We worked hard, but we had a good time, too. One Christmas we gave Mr. Denius a donkey. He was pretty stubborn sometimes, and difficult, and that's probably why we gave it to him. But probably more than any other reason, we had a chance to buy a cheap donkey. His children loved it, but the neighbors did not!

We had had an Aero Commander for quite a while, but I decided that if Tupperware could have a Lear Jet, and Frank Sinatra could have a Lear Jet, Radiation should have a Lear Jet. The Lear Jet really gave us a kick in the pants in growth. It was an inspiration to new employees. Where else could you go to work and fly on a Lear jet a couple of times a year, or every month if you were in sales?
— George Shaw

Remembering George Shaw
A visionary with a lifelong passion for learning and community
George S. Shaw, co-founder of Radiation, Inc., passed away June 15, 2010, at the age of 88.

In 1950, Shaw and his friend, Homer Denius, started Radiation in Melbourne, Florida. The company quickly became a leader in the design and manufacture of telemetry equipment for rockets. In 1967, Radiation merged with Harris Intertype to become Harris Corporation.

As the Space Age approached, Shaw visualized that the community and the world would soon change. His vision came partly from his discovery of printed circuit boards in a lab while attending the University of Florida and working in Washington, DC on top secret projects. He said one of the biggest influences on his vision was working with rocket scientist Wernher von Braun on how to get into space.

Retired Radiation and Harris Corporation employees who worked with Shaw said they believed in him from the beginning and remember him as the foundation of the company's electrical engineering strength and inventive culture. But it wasn't always easy to attract some of the best engineers in the country away from established technology centers.

Shaw became actively involved in the community to make that happen, helping to transform Melbourne from an agricultural community into the Space Coast's technological and commercial hub. He lent his support to initiatives in education, the arts, healthcare and banking, leading to the development of community institutions such as Brevard Engineering College (known today as Florida Tech) and Holmes Regional Medical Center.

The name Radiation, Inc. came from a set of engineering encyclopedias Shaw had in college. "It was all about radiating communications, yet after the Cold War years, the name needed to be changed," Shaw said.

After the merger that led to the name change in 1967, Shaw stayed at Harris until the end of 1968, when he and other Radiation executives, including Denius, resigned to devote their time to a new company they started, Electro-Science Management. This pursuit required less of his time, so Shaw soon went looking for new adventures, purchasing land in Costa Rica where he, along with his sons, helped create a new town, including a school for the children of those working at the town's sawmill, cattle ranch and coconut farm.

His lifelong passion was education, believing it was the answer to bringing peace between nations and better health to the world.

While Shaw lived away from Melbourne for 20 years, he visited the area frequently. In February 2010, he joined current and former company executives as well as community leaders in a celebration marking Harris' 50th anniversary of the opening of its Palm Bay campus.

"George Shaw looked at the world with a unique vision, always with an eye toward finding the new, exciting possibilities that existed around him, whether in science or in the community," said Dan Pearson, Harris Executive Vice President and Chief Operating Officer. "His vision helped shape both our company and Brevard County, along with everyone who had the privilege to meet or work with him. His creative spirit, humor and commitment to our community will truly be missed."

Memories of Joan Sherman

Joan was a chemist working on motion picture processing for RCA at Patrick AFB in the late 1950's where she met Dr. Jerome Keuper, who was trying to organize Brevard Engineering College, which eventually became Florida Tech. She taught several courses for him during the 1960's, including algebra and chemistry, in various temporary classrooms in churches and other buildings, including cabins on the site of the present campus.

In about 1962 she heard of an opening for a chemist at Radiation, Inc. She applied, but was told that they "were looking for a man." In a short while, they informed her they had a position reporting to a newly hired male chemist, which she accepted. It turned out that her "supervisor" was a young new-graduate, with much less experience than Joan. In a few weeks he quit for a government job and she was offered his position. She pointed out that she had previously been considered unqualified for the position. They told her that "they knew her better now."

She accepted the position and had a key roll in the new Micro-Electronics Division, formed to design and build integrated circuits, ultimately becoming a director. She recalls working under microscopes with with executives Joe Boyd and Jack Hartley to frantically inspect and encapsulate early IC's which were overdue to an important customer.

Joan continued working on integrated circuits as Harris acquired the semiconductor operations of GE, RCA and the original Intersil. In 1999 Harris spun of their entire semiconductor operation into a reborn Intersil. There are still semiconductor plants on the north side Palm Bay Road.

Joan is now retired, but still lives in the area.

Remembering Don Sorchych

Don Sorchych was employed in 1960 as an Engineer. He was promoted to Senior Engineer in 1961, Lead Engineer in 1962, and Section Head in 1964. He was appointed Director MED (Microelectronics) Product Development in 1966, Director MED Engineering in 1967, and VP of MED in 1968. 1979 he resigned and became CEO of Datamedix in Boca Raton. He founded and was CEO of Medicomp, a medical equipment company in Melbourne. He was also President and CEO of the microelectronic subsidiary of General Instrument Corporation in Phoenix

He is now Editor and Publisher of Sonoran News, a bi-weekly paper in the Phoenix area. He writes:
Recollections about Radiation/Harris
By Don Sorcych

My first job was with Radiation in 1960 after graduating with a BS in Electrical Engineering. Since I had been in the Navy for four years I was about 30.

1960 was the first year Radiation recruited the Big Ten schools and I quickly found out my fellow engineers thought I was there to integrate the schools, etc. My training was reading a pamphlet by circuit designer Bill Eddins about a coder. I asked engineers Bill Hicks and Bud Mills what a coder was. Hicks said, "Figure it out yourself, you fucking Yankee." We later became friends after they realized I had no political ambitions. In 1960 Florida was segregated – restaurants, movies, drinking fountains and housing. Radiation was hated by most of the community because as a government contractor the inspectors who came to evaluate us were Black. They wanted to eat at local restaurants, play at private golf courses and break down race barriers. It took years before we were accepted.

I was placed on a fix priced price job for Northrop as a circuit designer. Because the project was losing money, we were scheduled to work a 95 hour work week for 40 hours pay of $8,000 per year. Right down the hall a cost plus job for ATT to design a space craft to compete with Sputnick was paid straight time with overtime.

When the Northrop job was finished I was sent to California to install it. Fortunately Northrop knew they were getting a crappy system and signed off on it. When I returned I found my review gave me a three percent increase with a statement I would never be manager. My boss was a guy named Don Ellis who had just quit to join competitor Spacecraft Inc.; he managed to get away before I could kick ass. So I quit and the leader of an analog design group named Lonnie MacMillan asked me to join his group and he redressed my salary. I had a great year and loved my work. I was then appointed to lead a telemetry design group in the Aerospace Division working for Bill Vernon. We developed telemetry systems for the Minuteman Missile and the Apollo systems including the lunar lander. The waste with NASA was appalling because we built systems they demanded and after several iterations they went back to the first system we described in our initial proposal.

Joe Boyd was hired as the new Radiation President with a clear objective to sell the company. The founder, Homer Denius, was ready to cash in. Co-founder George Shaw stuck around for a few more years.

Joe had pressured me to go to what he had titled The Physical Electronics Division as the head of engineering. I liked what I was doing and had no confidence in the leader, Urey Davidsohn, nor the marketing guy they sent over to find out why so much money was being spent. I have an editorial in Sonoran News archives summarizing Urey's tragic story. I also have an archival story about Peter Petroff, the ex-Nazi soldier who convinced Shaw microelectronics was simply a way to get smaller systems. Shaw's dream was to get a small cheap chip that would be an A-D converter to instrument process manufacturers rather than the analog systems in current use. He used to come into my office, put his wallet on my desk and promised $100,000 if I could make one. I spent hours explaining why we were not even close to the desired chip sizes.

Boyd arranged an employee meeting and introduced Harris Corporation Chairman George Dively. There was a Fortune magazine article which thusly described Dively: "under his gruff exterior he has a gruffer interior." After pleasantries Boyd asked for questions. Silence. Boyd said, "A.B., you always have questions." A.B Amos said "I do. As I understand it Harris is profitable, while Radiation is a growth company. So we will provide the growth and Harris will provide the profit." Dively was stunned and after a moment he said, "First of all, are you serious?" Amos said he was. Dively launched into a bitter diatribe, suggesting there are two kinds of brains; the one who says profit and growth go hand in hand and the other who offered a stupid suggestion. No more questions, just silence.

Nonetheless Harris bought Radiation in 1967.

In a short while I was promoted the General Manager of Physical Electronics and soon we changed the name to Harris Semiconductor. Radiation was always proud of its red logo and Harris adopted a version of it. Harris Corporation consolidated corporate headquarters in Cleveland and Boyd moved there. Harris had counterparts of major functional roles, such as VPs of Manufacturing, Human Resources, Engineering, Research, etc. They believed in regular visits and there were soon problems. Their head of manufacturing was headstrong guy who visited and began to order changes. I kicked him out and got a visit from Harris Corp. President, Richard (Dick) Tullis. I explained we were a decentralized company and corporate was an annoyance with no knowledge of what a semiconductor was. The Corporate Director of Research called Texas Instrument demanding royalty payments for few puny patents we had and forced us to negotiate an expensive cross license with TI, which helped take the pressure off. I wrote an obit about Tullis when he passed. I had nothing but respect for him and we worked closely over the years.

When Dively retired, Tullis became Chairman and Boyd became president. Boyd decided to bring Harris corporate to Melbourne. Tullis and I fought the idea for years. I got a call from Tullis saying, "We lost; the board ruled against us, so I will move to Melbourne and assure we are a friendly corporation in how we handle the community." Tullis moved to John's Island near Vero Beach but was active in community relations.

Semiconductor had myriad problems. Internally, systems division wanted Semiconductor to be their captive supplier, which made no sense. In the meanwhile ex-TI engineers were sabotaging experiments of one engineer they didn't like. So I fired the saboteurs. All of the product was coming out badly inverted so Boyd hired a process scientist from the University of Michigan – that didn't help. Engineers didn't listen to a solution technician Ed Guerra found in an IBM technical paper, which was the use of phosphorus glass on planar surfaces. It solved the problem.

Semiconductor had an unusual technology called dielectric isolation, brought there by Davidsohn but developed and improved by John Short. That technology allowed broad bandwidth linear products and also radiation hard products. We competed effectively and won Polaris and Poseidon programs. In those cases we were working with MIT and Raytheon.

Our organization expanded to Government Programs, Linear products, Digital products, R&D, Production and Materials. We were also the first in a CMOS product line.

At that point I was promoted to Vice President & Group Executive.

I have mentioned Joe Boyd several times. There is much I know and much I could say, but let me be positive: Given every thing about him he was a highly successful and effective executive. In all our years together he never directed me and we always found ways to get along. Both Radiation and Harris were planful companies. We did five year plans, annual plans and quarterly plans with thorough reviews. We did annual personnel reviews and we had monthly corporate meetings.

I had made it clear who I would work for and who I wouldn't to both Boyd and Tullis. When Tullis was retiring, he called to tell me I had missed the president's position by one vote. So in 1979 I resigned and became CEO of Datamedix in Boca Raton.

Carl Swisher Remembers

I imagine everybody has a story to tell about their history at Rad Inc. AB's was enjoyable for me to read because he mentions so many familiar names.

Of special interest to me was his mention of Bill Dodgson and his philosophy of marketing. I first met Bill when I was working at Rocketdyne's Neosho, Mo. Facility in 1960 when Bill and Perry Knight paid us a marketing visit trying to sell us on SEL's advantage of using their low level multiplexor.

Rad Inc. had been there first but their $1000 per channel approach priced them out of the ballpark. For some reason, which I never fully understood, SEL offered me a job as a logic designer. I would not have known what a flip flop was if it jumped up and bit me in the ass but with the help of Dave Yoder and Larry Klinger I learned fast.

Dodgson's philosophy of marketing was referred to as B & B marketing, "Booze and Bullshit". But with the success of the low level multiplexor and Dodgson's marketing, SEL quickly captured a significant segment of the Data Acquisition Market.

The original group at SEL consisted of former Radiators. Bill Dodgson, Tommy Thompson, Perry Knight, Larry Klinger, Gus Randolph, Lonnie McMillian, Ed Clagget, Buddy Mock, Clay Myers and Bill Crowder.

They later added RB McPhail, Dave Yoder, John Searcy, Stan DeShazo and Guy Bradford. John Searcy and Guy Bradford were the circuit designers that were primarily responsible for the success of the low level multiplexor. To the best of my knowledge I was the first non-Radiator engineer hired.

After a few months of internship I was given the task of designing a multi-channel system for NASA Huntsville. When it was delivered NASA was in the process of starting check out the multi-stages of the Saturn Launch Vehicle. The company that had the contract for the Saturn checkout system did not make delivery and I modified the SEL system to checkout the Saturn, one stage at a time. The multi-stages were successfully checked out on time and the other company's contract was canceled. NASA gave me a contract for 6 more systems to hand carry back to Ft Lauderdale, along with a written specification for a computer that was disguised as a Data Acquisition System because only NASA's Accounting Dept was authorized computers.

This contract from NASA was SEL's introduction into their successful computer business. Since I felt partially responsible for this contract I asked Perry Knight, who was VP Engineering, if I could work on that contract. He declined and insisted that I be the Project Engineer for the other 6 systems. I felt that this was not fair since it would not offer me any opportunity to grow, only to make money for SEL.

After a 2-hour closed-door session Perry and I decided to part company and he suggested I try Rad Inc., which I did, and the rest is history.

The former Rad Inc engineers used to tell a story about taking Bill Dodgson bear hunting with them in Alaska. The all agreed to meet at the Melbourne Airport and let Roby Green fly them to Alaska.

They all met there on time with all of their guns and ammo, etc. After they had everything loaded on the airplane and were about ready to take off a taxi came roaring up and stopped by the airplane. The door opened and Dodgson spilled out. They drug him into the plane and took off. When they arrived at the cabin Dodgson hit the sack and they all stayed up all night drinking beer and playing poker.

When morning came they were all sacked out but Dodgson woke up all bright eyed and bushy tailed and opened the door and took off into the woods with no gun. After about an hour they heard him screaming and went to the window and there he was running towards the cabin with a bear right on his tail. They opened the door for him but he side stepped at the door, pushed the bear on in, slammed the door and hollered,

"You guys skin that one while I go get another one."

Any engineer that worked on a contract that Dodgson (or any other marketer) sold understands the meaning of that story.

Ted Woodward Remembers

From a Letter

Dear Frank,

Enjoyed talking with you today.

I'm Ted Woodward
2506 Forest Dr.
Melbourne 32901

As I mentioned, I'm Johnnie Hiott's brother-in-law. My wife Laura is his sister. Honest John's fish camp is the site of Laura's grandfather's homestead. He and a brother homesteaded 160 acres there during the Teddy Roosevelt administration (1885-86). Laura was born in Grant in 1930 and graduated from Mel High in 1947 (the old Henegar buildings). She is a good source of Melb. info should you need a resource.

As I mentioned I'm sending you:

A large photo of the 'RadiQuad' antenna from 1958. Seven were designed, built, and placed around the world in order that Eisenhower could transmit a Christmas message from space. The quad helix antenna was put on WWII searchlight mounts.

A photo of an antenna test facility prior to our move to Palm Bay. The persons in the photo are Marv Livingston, Leroy Nash, & Stu Henderson.

Another photo of the test antenna.

A small photo of antenna feed. This project was to design, develop, fabricate, test and ship the antenna feed for the English Jodrell Bank facility. As I recall, one was for a 60' dish and one was for the 250' radio telescope. Radiation did it! I can recall one of our drafting people Norma Rockefeller, was sketching the plumbing to make drawings as the feed was being painted for shipment!!

An invitation to the first Radiation Old Timers Day in 1996 at A.B. Amis home in Grant. Lots of names etc.

A list of Harris ESD long service employees in 1997. Lots of names etc.

I also mentioned that Johnnie Hiott had brought me a copy of your Chapter 9 for your book. I really enjoyed it. What memories!!

In addition, I recall the Castaways Restaurant on the bay in Palm Bay. I recall that we ate in rocking chairs and that the menus were printed out on paddles (larger than ping pong). I think it burned down abou

Chapter 4

Plants and Facilities

Radiation began in World War II-era buildings at the Melbourne airport. Melbourne had airports in several locations in the very early days, but he present location of originated in 1933 when the city acquired 160 acres west of the Indian River Bluff area, to replace an earlier airport off US 192 near the present Melbourne Square Mall. The current airport property was the site of a Naval Air Station during World War II, training over 2200 pilots in Grumman F4F Wildcats and F6F Hellcats. The property was turned over to the city of Melbourne in 1946, as surplus property. At that time there were 129 buildings at the airport.

Radiation occupied several buildings at the airport, including Plants 1, 2, 7, and 9. A 1961 aerial photo locates these buildings. Plant 1 was the first Radiation locations, and Plant 2 was leased in 1953. In this period a qualification for Radiation engineers was was rumered to be that they were able to paint.

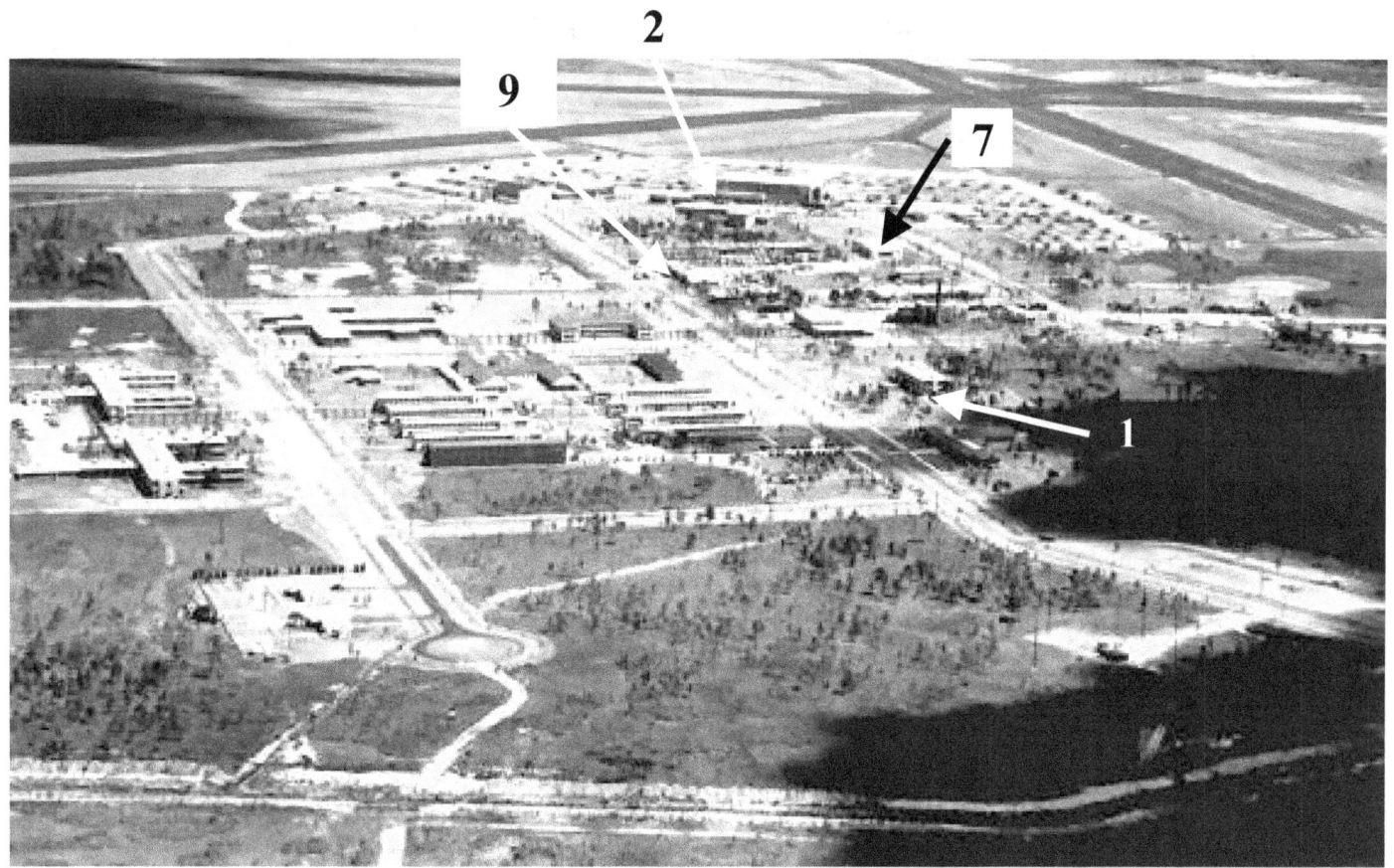

A photo from the Naval Air Station days, the mid 1940's. The locations of the to-be Radiation buildings are indicated. The present day terminal is located near the large hangar near Building 2.

PLANT 7, MELBOURNE

1960 aerial photo of Melbourne airport with Radiation buildings

Melbourne area buildings. In November 1954, Radiation occupied two buildings on the Melbourne Airport, Plant 1, above left, and Plant 2, above right..

Plant 16 was the home of Publications.

Plants 5 (left) and 5A were on a remote part of the airport property, called "80 acres." They were used for reflectivity and antenna pattern measurements.

Plant 7 (above), 8(below left and 9 (below right) were located at the Melbourne Airport

Plant 6 was located on the Malabar Auxiliary Airport property and had a collection of tracking radars used for various flight tests.

Different publications identify this as Plant 3. Different angles, different buildings?

Plant 4, with the telemetry van for Operation Teapot.

By November 1954, Radiation occupied two plants, #3 and #4, in Orlando, near Herndon Airport. One reason for this was to have access to the longer runways in Orlando, for work on instrumenting jet aircraft for Operation Teapot nuclear weapons tests. Access to the larger labor pool in Orlando may have also been a factor; a production facility was established in Plant 3, under Dean Ashwell, one of the earliest Radiation employees. Eventually the Instrumentation Division was located briefly at 440 Boone St., and later moved to a more modern building near McCoy AFB (Pinecastle), site of the modern Orlando International Airport, and was renamed Systems Development Division–Orlando. There was also a Research Division in Orlando, occupying part of Plant 4 and later a building on Hermann Ave.

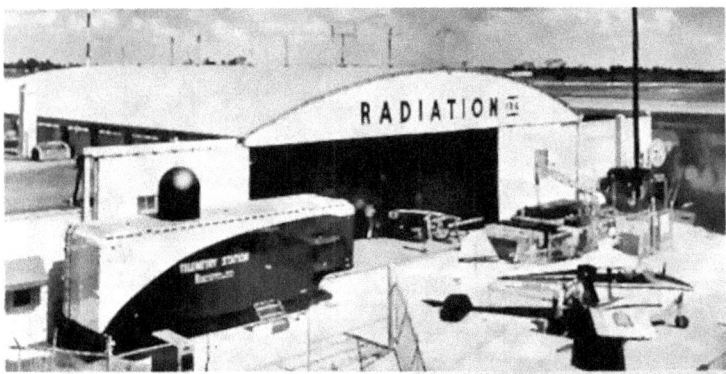

"New Instrumentation Division" plant.

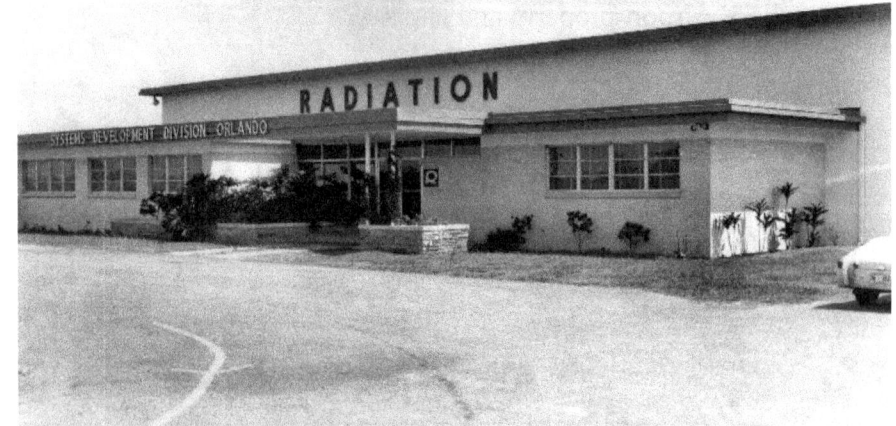

Radiation felt the need to establish a presence in California, to support our customers there. In May of 1959 the Space Communications Division was established in a 12000 square foot facility in Mountain View(left). They were known for a showy Mission Control room for the west coast test range.

In 1960 Radiation acquired Levinthal Electronics Inc, in Palo Alto (above). Levinthal built high power transmitters, including a 430 MHz radar transmitter at Arecibo Observatory in Puerto Rico producing 2.5 MW of peak power.

Levinthal eventually became Radiation-at-Stanford,(left) and built several types of systems, including the Radatac data acquisition cart.

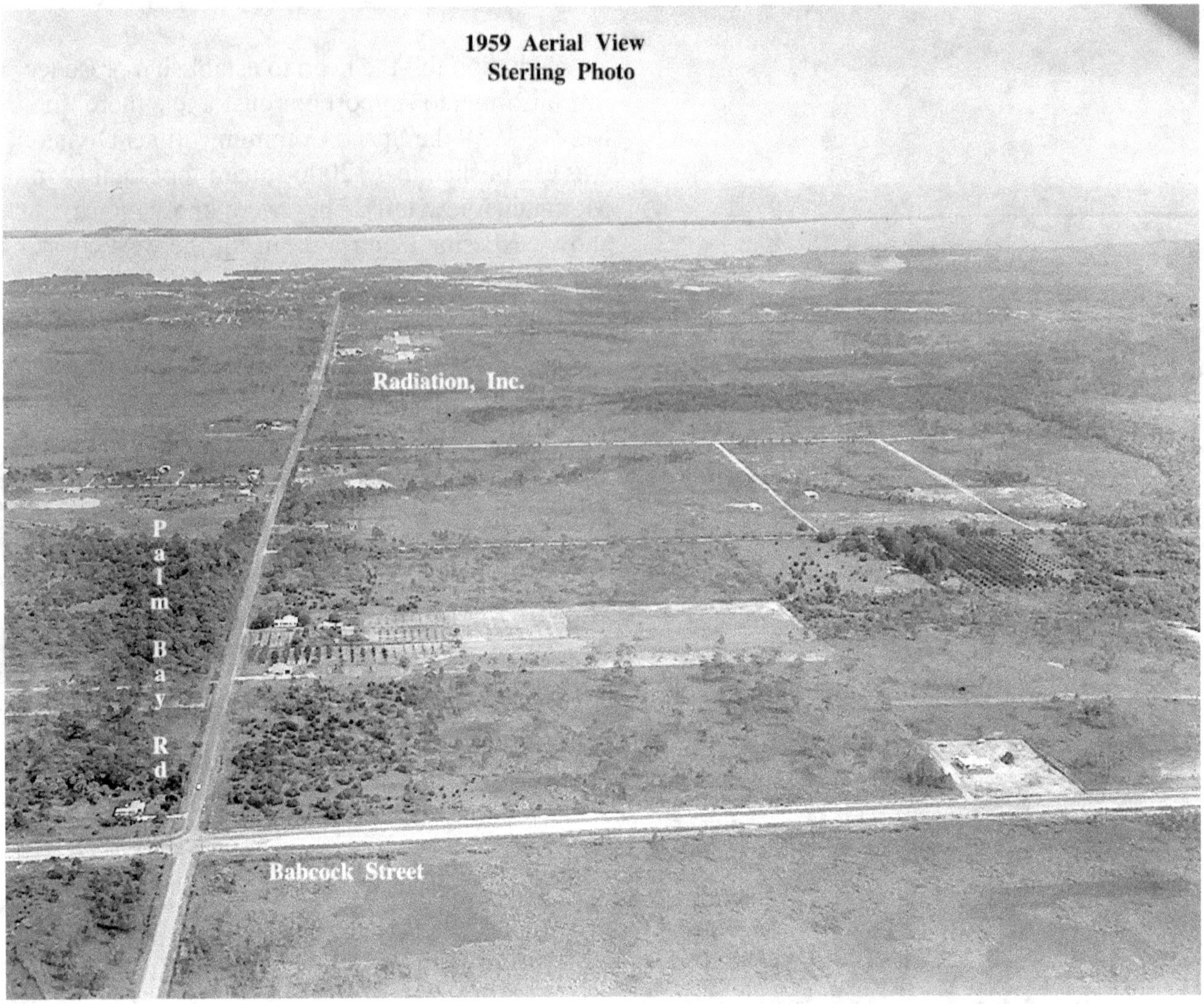
1959 Aerial View
Sterling Photo
Radiation, Inc.
Palm Bay Rd
Babcock Street

In 1959, Radiation began moving to a new campus in Palm Bay, quite a remote area in those days. Below are the first three Palm Bay buildings under construction.

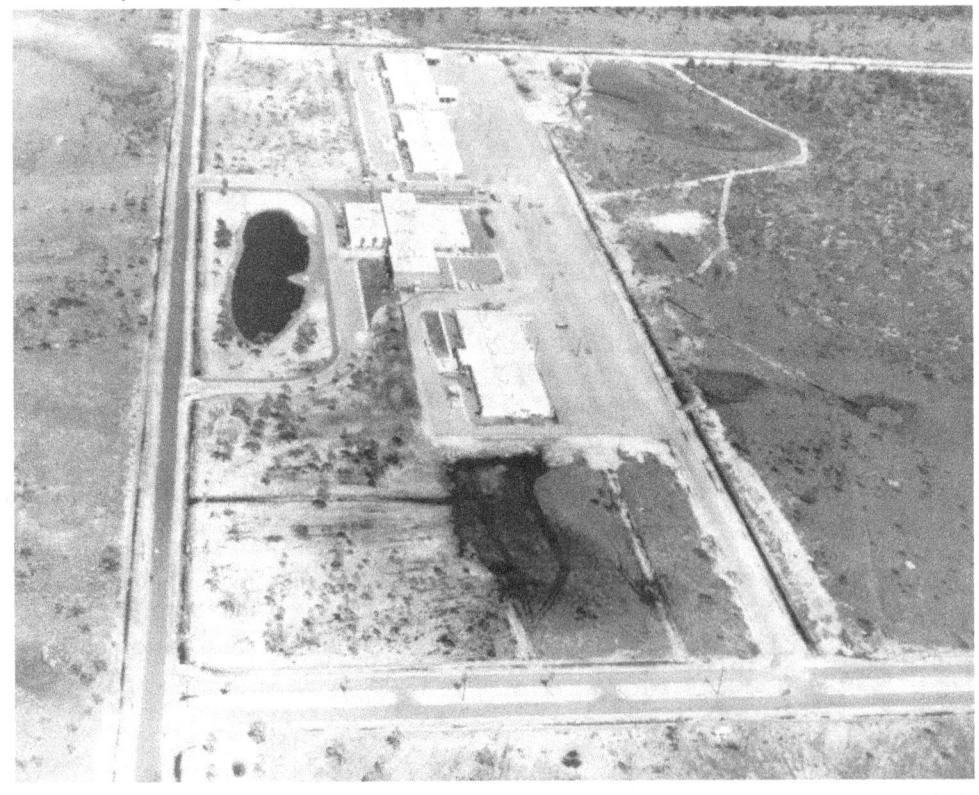

Above, shortly after the first employees moved in, late 1959 or early 1960. Note that what is now building 1 was originally two separate buildings, called B and C.

A few years later, several buildings have been added, notably 11 and 6A.

A 1966 plan of the campus, showing how small the original buildings loom in the modern layout.

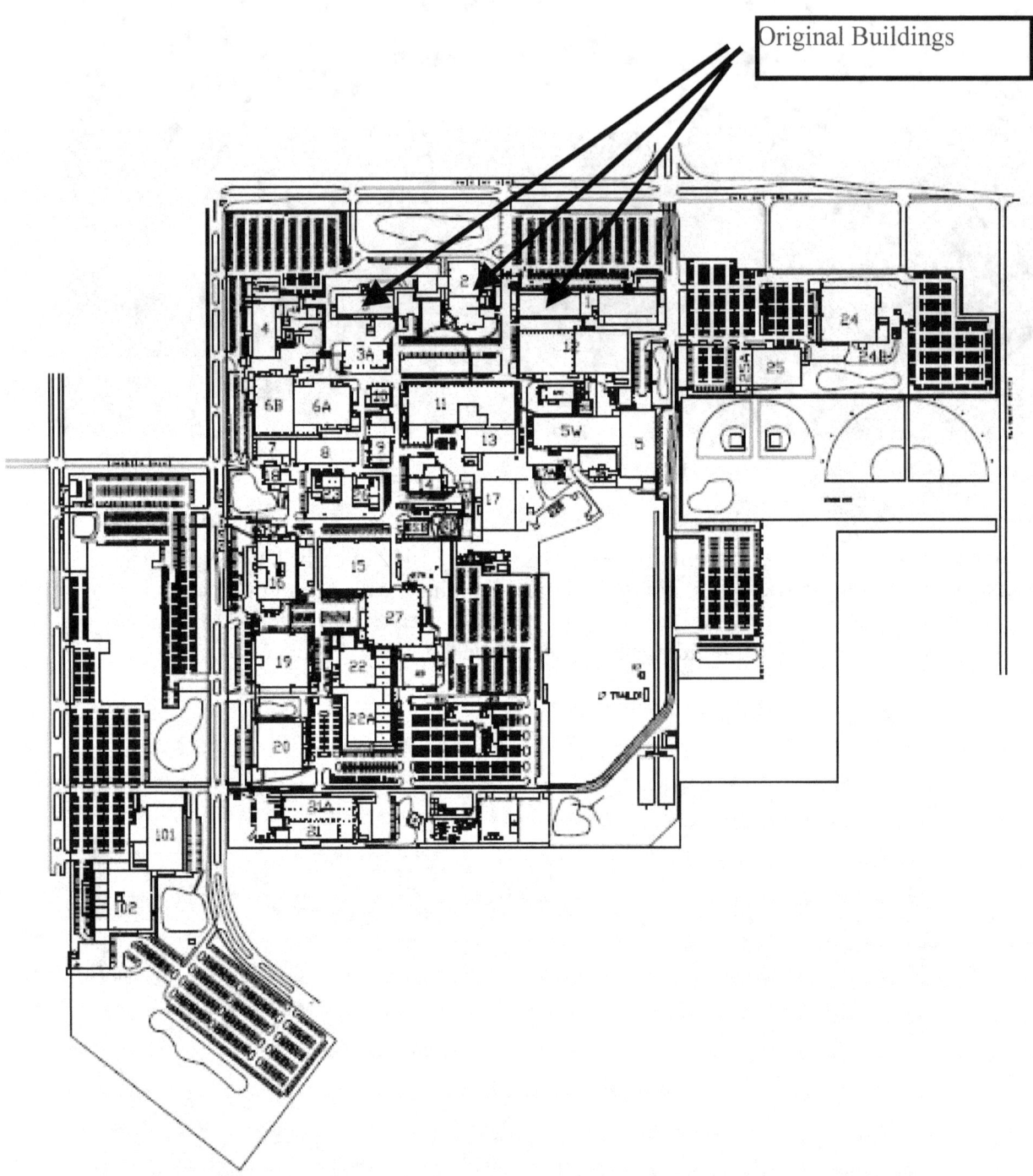

Chapter 5

Major Programs, Products and Systems

A description of some of the programs at Radiation gives a taste of the variety and challenges involved. Some of the jobs reflected the interests and knowledge of the founders and early employees, but serendipity also played a large role. You had to find a customer who believed in your ability to solve his problem. The founders believed that the more accurate measurement of information and the faster transmission of the information to where it was needed, mostly through digital techniques, was the key to success but this still left a broad range of challenges, as we will see.

Telemetry Systems

The earliest Radiation digital (PCM) telemetry system was the AKT-14 airborne system and UKR-7 ground station. Radiation received a contract from WADC (Wright Air Development Center) in about 1955 to salvage what they could from an earlier prototype built by Melpar. **Ruggedized and improved versions of the AKT-14 PCM (Pulse Code Modulation) system were built for Convair Division of General Dynamics and Avco Corporation.** The General Dynamics systems were apparently intended for WS-107, the Atlas ICBM (InterContinental Ballsitic Missile) program, although I find no records of PCM actually being used aboard Atlas. The Avco systems were for their work on re-entry nose cones for Atlas, Titan or both. A modified version of the AKT-14 was built for the Flight Test Lab at WPAFB, recording the data on an on-board tape recorder instead of transmitting it to the ground.

From 1959 through 1966, Radiation built a series of PCM systems for at least 11 customers and programs, including one aircraft program (Norair F-5/T-38), one jet engine program (GE), two missile programs (Titan II and Minuteman), four unmanned satellite programs (Telstar, Lunar Orbiter, Nimbus, Orbiting Astronomical Observatory), the Apollo Command and Lunar Modules, and a system for the supersonic rocket sled at Holloman AFB.

Another early Radiation telemetry programs was the airborne AKT-6 and the accompanying ground station, the UKR-1. The AKT-6 was a Pulse time Modulation system; the information to be transmitted modulated the time or position at which a pulse was transmitted. The pulses were transmitted at S-Band, around 2200 MHz, a frequency little used for communications in those days (early tracking radars also operated at S-Band.) In the literature, mention is made of the 1947 development of the AKT-6, perhaps for V-2 rocket tests. Radiation's involvement with the AKT-6 was prior to 1955, when it was first used in Operation Teapot. Radiation performed significant redesign and modifications to the earlier AKT-6.

Instrumentation Systems

The field of instrumentation was another area of interest to the founders.

Operation Teapot

An early and significant program was the the instrumentation of unpiloted aircraft (drones) for operation Teapot, a series of tests of atomic warheads in the Nevada desert, in 1955. The program was carried out by the Instrumentation Division in Orlando (for access to a long runway for the jet aircraft involved).

A company brochure of the period states:

"In one of its most comprehensive projects, Operation Teapot, over 340,000 man-hours were applied by this division in design, installation, field operation and maintenance of a multi-unit telemetry system, modification and instrumentation of four jet aircraft, equipping and operation of mobile ground stations and evaluation and reporting of test results."

The jet aircraft mentioned were QF-80's, a drone version of the Lockheed F 80 fighter.

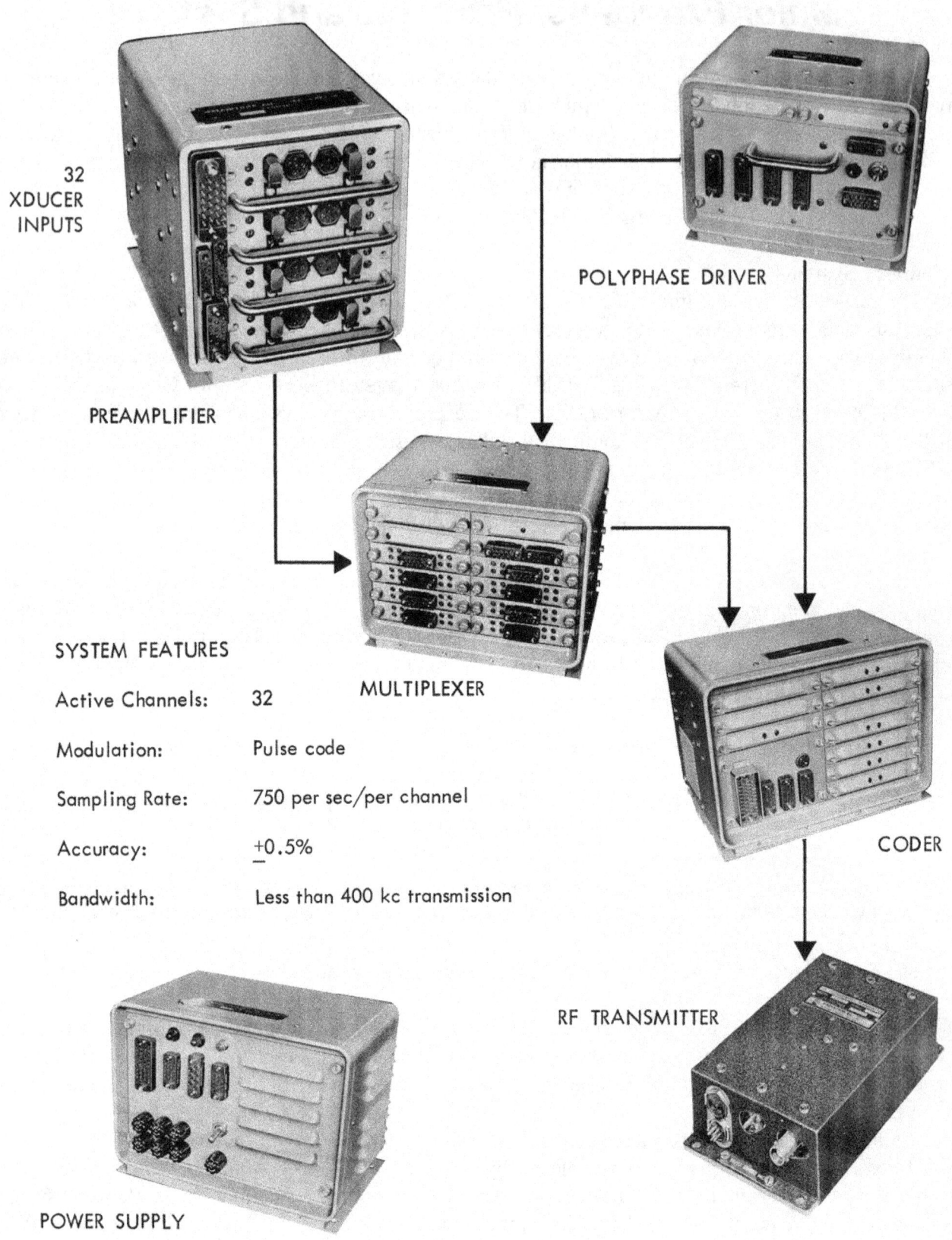

Norair PCM

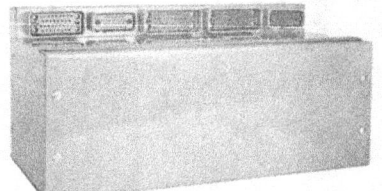

MULTIPLEXER

MISSILEBORNE PCM SYSTEM DEVELOPED
FOR USE WITH THE TITAN MISSILE

SYSTEM FEATURES

Input Signals:	64 analog
	40 bi-level
	1 computer
Signal Level:	± 5V
Programming:	Variable
Word:	8-bit
Weight:	19 lbs
Volume:	Less than 1/3 cubic foot

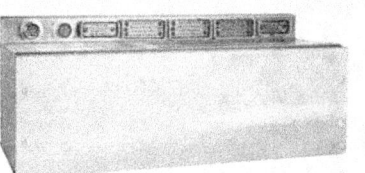

PROGRAMMER

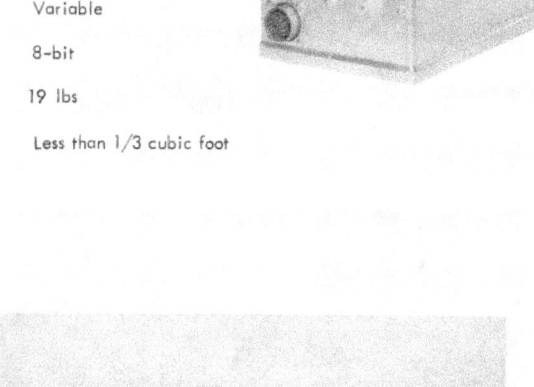

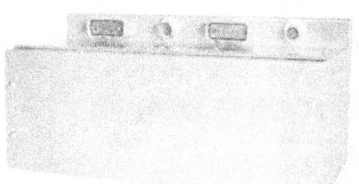

ACCUMULATOR

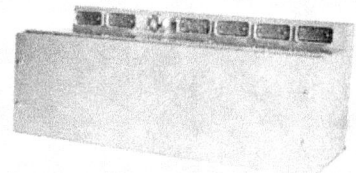

CODER

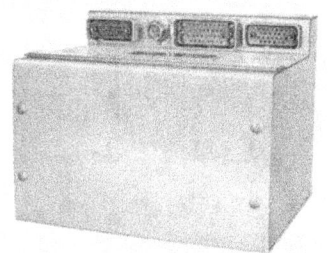

AMPLIFIER

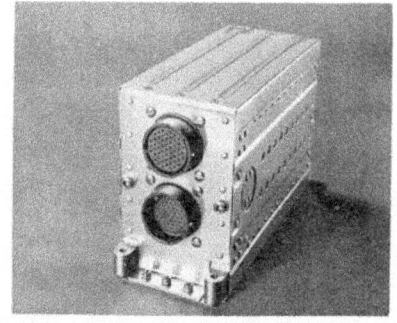

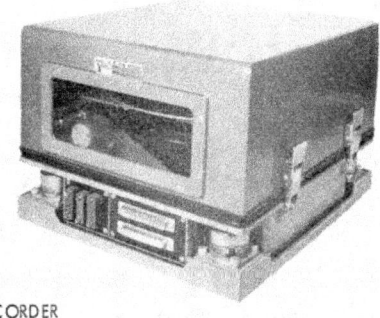

RECORDER

Minuteman PCM

Apollo Command Module PCM System

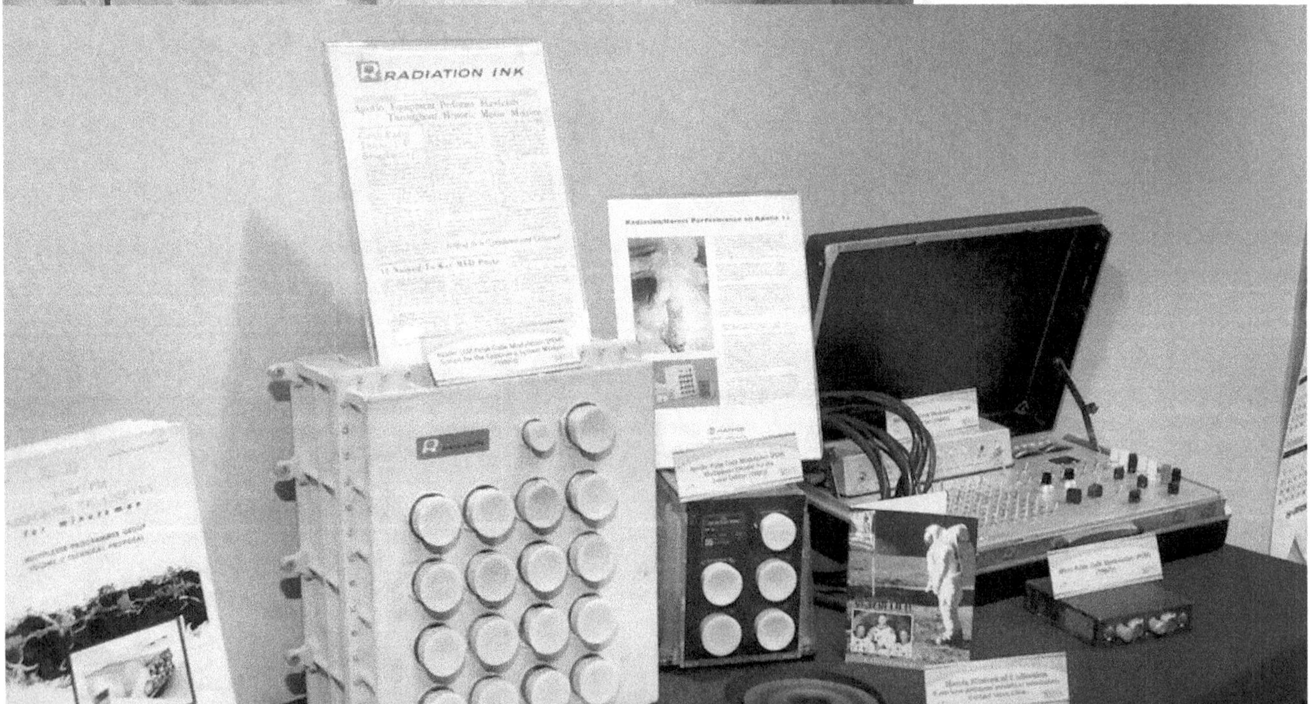

LEM (Lunar Excursion Module) and Lunar Orbiter PCM Systems

One of the QF-80 drones instrumented for Operation Teapot

The ground station for Teapot

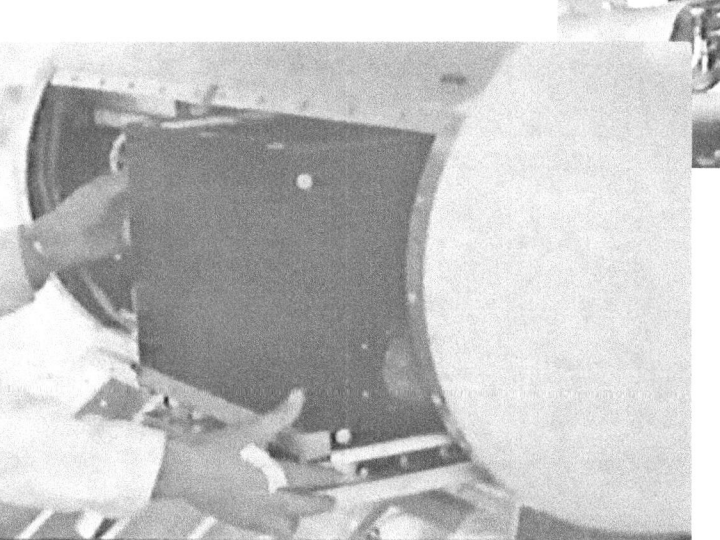

Recording camera in QF-80

Instrumentation in QF-80

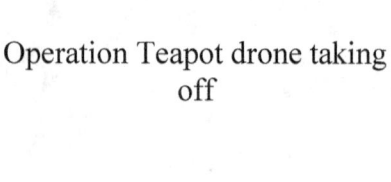
Operation Teapot drone taking off

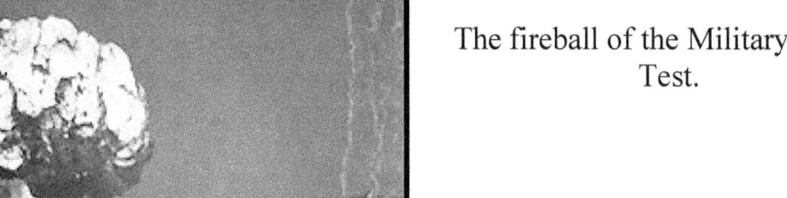

Inflight. Drone at center, control aircraft at top. In foreground is an armed fighter to shoot down out-of control drones.

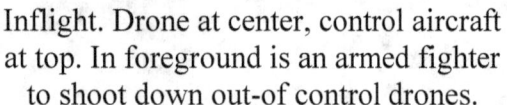

The fireball of the Military Effects Test.

Of the three drones, one was destroyed by the blast, the second survived with damage, but control was lost. The third survived but collapsed a nose wheel on landing.

The drones were used in the "MET" (Military Effects Test) shot, a 22 kiloton blast on a 400 foot tower, conducted on April 15, 1955. Of the four instrumented drones, one was lost to a control failure on takeoff, but the other three were successfully flown over the blast at altitudes of 3800, 4300 and 5100 feet over the tower top, to assess the damage at different ranges from the blast. The lowest drone was, as anticipated, significantly damaged and crashed soon after the explosion. The medium altitude drone was less severely damaged, but during an emergency landing attempt, control was lost and the plane crashed. The high altitude drone sustained only minor damage and was flown back to Indian Springs AFB and landed. Unfortunately the nose wheel collapsed on roll-out and the drone ended up in the desert.

Force Amplifier/Indicator

Another aircraft instrumentation job was the Stick Force Measurement System. Hardware from this job, the "Force Amplifier" is the earliest Radiation hardware known to still exist. Several examples of which were purchased on E-Bay by Nancye Meyers and Tim Hartsfield, Radiation memorabilia collectors at Harris Corporation. The amplifier is pictured in a Radiation brochure dating to about 1955. The company brochure states "Radiation Inc... developed and is producing a Servo Force Measuring System This equipment was designed to measure and indicate directly the forces applied by the pilot to the various control surfaces of an aircraft in flight. The indicator is house in a standard instrument case and is normally installed in a photo-recorder panel where readings are photographed as desired."

A later paragraph in the same brochure probably refers to the same equipment:

"...design and production of the S-6 servometer system for the flight test of aircraft. The complete system provides accurate dial indication of tension and compression forces as applied to aircraft rudder pedals, and wheels or stick assemblies. These highly accurate servo amplifiers may be used with many different types of pickups and transducers."

The Force Amplifier is also notable because it uses vacuum tubes, specifically:
a 6U8 triode-pentode
a 12AX7 dual triode
a 6AQ5 power pentode

The earliest contract was AF04(611)-424 with Edwards AFB and was assigned JA 1040 under Ralph Johnson. Other systems were built for the Navy (P.O. 1-2278-55, JA1052) and North American Aviation (P.O. H-562-A-33566, JA 1053), presumably for installation in other aircraft. These JA's were assigned to Fred Cullman in Orlando.

The most challenging aspect of these programs was installation of the sensors and other equipment in the existing aircraft, with minimum modifications..

Ground Data Systems

A big variety of digital ground systems were built at Radiation. Some served as receiving stations for airborne systems like the the UKR-1 and UKR-7 mentioned above. Many served as interface devices for general purpose digital computers, which were becoming popular but which were still geared to processing static financial data presented on punched paper cards, and were not friendly to large quantities of scientific or engineering data. The usual approach in the ground stations was to generate computer-compatible magnetic tape which the computers could absorb and process at their own pace.

The Radiation ground stations were usually either data acquisition systems which digitized local measurements, or receiving stations which accepted data from remote telemetry systems. Data was displayed and recorded in a variety of forms, analog and digital, such as meters, chart recorders and printers. Other ground stations accepted data from a computer and formatted it for transmission, as for satellite commands. More specialized ground stations will be mentioned in other sections.

Ground stations varied in size from one rack, to some with dozens of racks and consoles, and were

Force Amplifier and associated display.

Below is an interior view, showing the vaccum tubes.

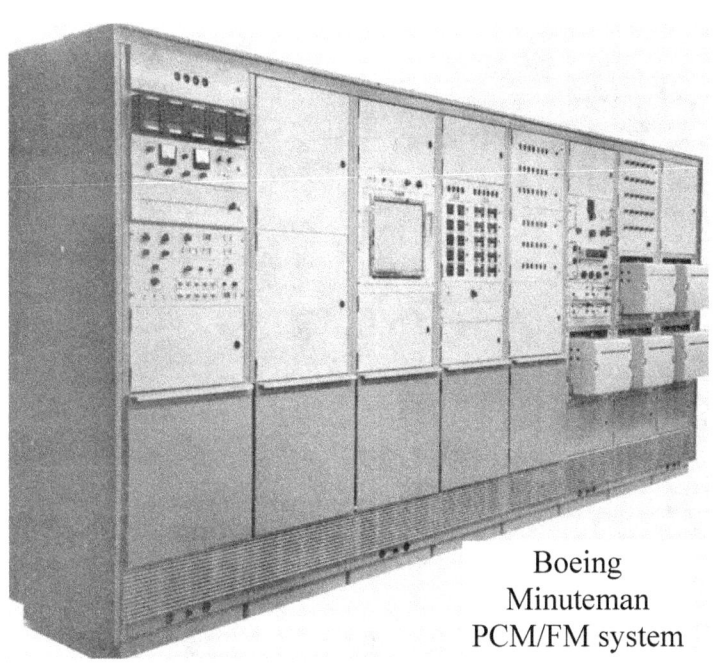

Boeing
Minuteman
PCM/FM system

PCM Systmem for
McDonnel Aircraft

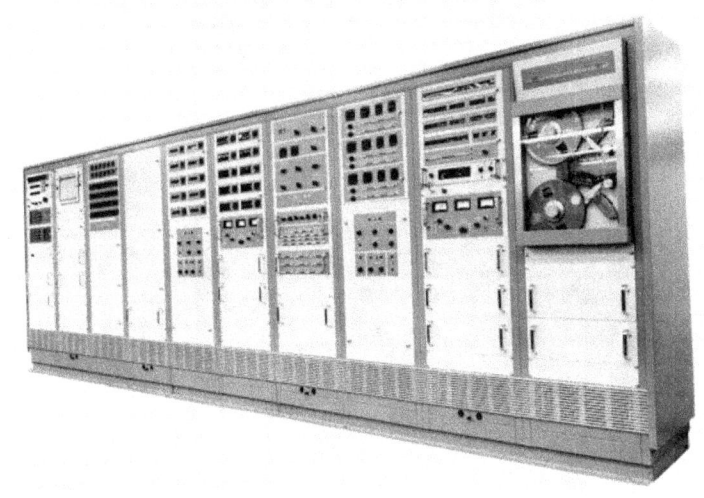

Nike-Zeus system for Bell Labs

Model 540 Decom for
NASA

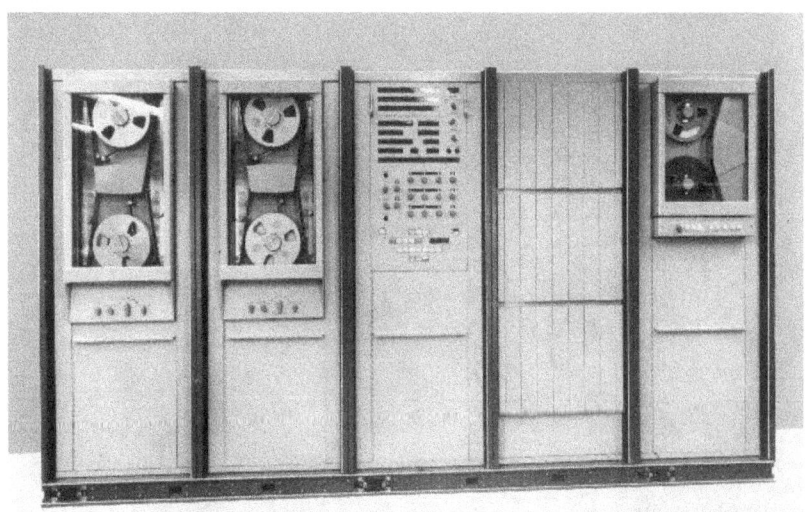

Data Processor for GE at Edwards AFB

Data Acquisition system for Thiokol

Gemeni DCS

Data Conversion Facility for Holloman AFB

built for a variety of customers and locations, including:
- Aberdeen Proving Grounds
- AC Spark Plug Missile Test Facilities
- Applied Physics Lab of the Johns Hopkins University
- AVCO Corporation
- Bell Telephone Laboratories
- Boeing - Minuteman Test Facilities
- Ford Motor Co.
- G.E. Edwards Flight Test Facility
- GE Missile and Space Vehicle Facilities
- Glenn L. Martin Company (Baltimore and Orlando)
- Holloman Air Development Center
- Kirtland Special Weapons Center
- Lockheed Missile and Space Test Facilities
- McDonnell Aircraft Corporation
- NASA (Goddard SFC, Johnson SFC, Kennedy Space Center, Marshall SFC, others)
- Naval Air Development Center
- Norair Flight Test Facilities
- North American Aviation
- Patrick AFB - Missile Test Center
- Thiokol - Rocket Engine Test Facilities
- USAF (Many sites)
- Vandenberg AFB
- White Sands Missile Range

Antennas

Radiation was a leader in large tracking antenna. They got into the business with an alternate proposal to the Eastern Test Range, who had intended to buy a dozen or so smaller telemetry antenna to outfit their down range stations. Radiation pointed out that it would be more economical to buy fewer, larger antenna which would provide the same coverage, and received a contract in 1956 for three TLM-18 60 foot (18 meter) diameter parabolic tracking antennas. Previous antenna of this size had been used for radio astronomy; the TLM-18 was the first capable of tracking high speed missiles and rockets. The first three were installed at Cape Canaveral, Antigua and Ascension Island. A different version, called R-1162, was installed on the Pacific Missile Range at Vandenberg AFB, and Kaena Point Hawaii. The Air Force bought another for South Point, Hawaii, and yet another was installed at Belmar, NJ. The New Jersey antenna is still in place at last notice. The 60 footer at Vadenberg AFB was installed in April of 1959. Another was installed at Hawaii in October, 1959.

An 85 footer was delivered to the AF in 1963, and another was used to beam TV from the Tokyo Olympics in 1964 to the US.

Another 85 footer, plus a 150 footer were delivered to Asmira, Ethiopia to monitor Soviet Space probes. This system is discussed later, under "Black" programs.

A 150 footer configured as a radar called Altair was delivered to **Kwajalein Atoll**. It is still in active use today.

The TLM-18 was the original large tracking antenna. Two were installed near Cape Canaveral, one at Antigua and one on Ascension Island., plus variants at other locations.

Antenna for Courier, worlds first active communications satellite

Left, 85 footer at Goldstone, CA.

Above: Altair is still in use on Kwajalein Atoll

ALTAIR 150 FOOT ANTENNA

Bayhouse (background) and Stonehouse Antenna in Ethopia.

Communications satellite antenna designed for roof-top installation, for a classified customer.

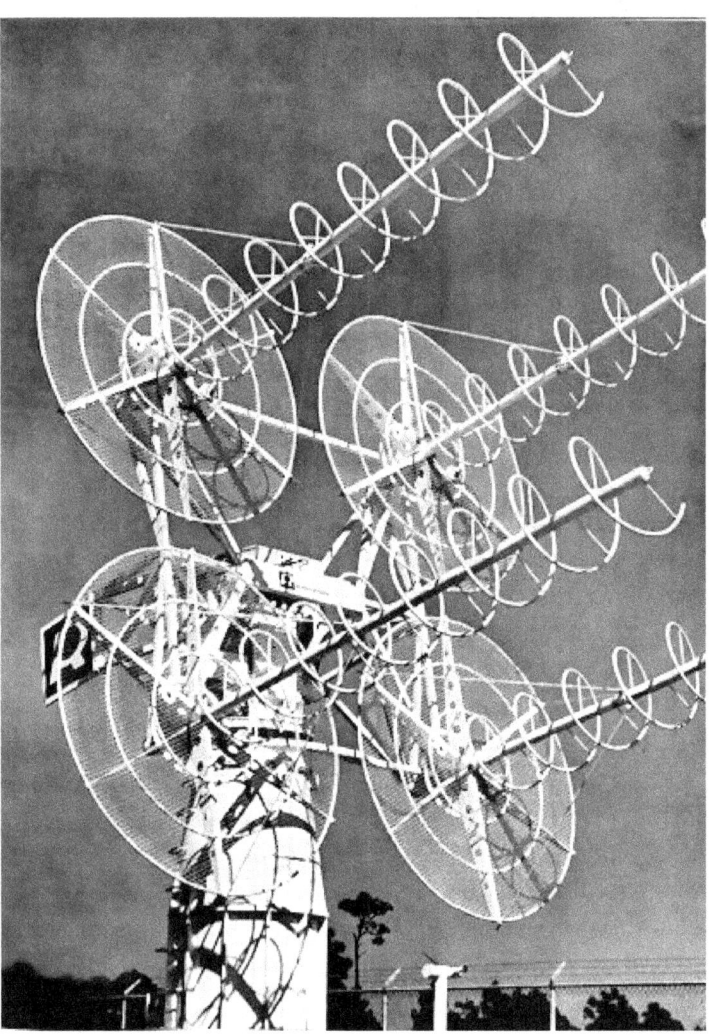

Quad helix antenna use for communicating with Telstar communications satellite.

Tracking antenna for USAF Dynasoar Program.

Black Programs

Some of the more interesting programs were the so-called "Black" programs, those requiring special level of clearance above those of normal DOD clearances. They were conducted in closed-off section of the plant and were insulated from many of the normal management channels. Usually they carried a rush schedule and challenging technology.

1300 (Hexagon) Program

In June of 1971, the first of a series of new photo reconnaissance satellites, called Hexagon, was launched into space. Shrouded in secrecy for 40 years, details of the system were released in 2011, and a satellite was displayed (for one day) at the Smithsonian. Little known, even locally, is the important contribution of Radiation to this program.

Details of the program are staggering: the satellite was 60 feet long and 10 feet in diameter (the size of a Greyhound bus) and included a telescope (called KH-9) of 60 inch focal length, capable of resolving objects smaller than 2 feet from the orbital altitude of nearly 100 miles. Each of the two large cameras carried nearly 30 miles (!) of 6 inch wide film which was returned to earth for processing in four re-entry vehicles designed to be snatched from the air as they parachuted to earth.

Hexagon was significant in that it allowed the US to relax its insistence on physical verification and agree to the Strategic Arms Limitation treaty.

Bob Landis of NASA said "Frankly, I think that HEXAGON helped prevent World War III." , and "This is still the most complicated system we've ever put into orbit ...period."

Radiation's role in Hexagon, or project "1300" as it was known internally, was as subcontractor to Perkin-Elmer, who developed the camera system. Radiation's initial task was to package the electronics, but ultimately had responsibility for the design of all the electronics.

Scott Broadway, who had a prominent role in the program, writes:

"Recent news has highlighted a program that was very important to the Nation, Radiation and Harris, but most importantly, for many of the people who worked on it. This program was highly classified and only a few of us knew the full nature and scope of the program while we were working on it, so now that it has been declassified, this is a good opportunity to let you in on the secret.

"The full details of the program can be found by Googling the terms KH-9 and Hexagon with the reference to "Spy Satellite". A brief overview of the program concepts will be covered today. This is especially for those of you heroes who did not know what was going on even though you may have worked on the 1300 Program. It is very important that you understand why we worked such long hours under such strange conditions. You really were heroes in a national security sense, because the results of this program made the negotiations with the Russians real and led to the success of the Strategic Arms Limitations Talks or SALT. We did not know that at the time, but 45 years later you can talk about it and I guarantee your "kids" will be impressed. Mine were!

"The basic function of this program was to get high resolution photographs over a wide swath via cameras on board a very large vehicle orbiting at a low altitude, and bring these images back to earth for interpretation. The images were returned to earth as rolls of exposed film via four re-entry vehicles recovered, in mid-air, by airplanes over the Pacific Ocean. This sounds complicated and it was. The payload itself was so large it required a very large satellite structure (60 ft. long by 10 ft. diameter, about the size of a Greyhound bus) and a Titan IIID for a launch vehicle. Lockheed was the total system integrator and many other organizations provided much of the program functions under the umbrella of the National Reconnaissance Office with the CIA, Perkin-Elmer Corporation, the Air Force, JPL, GE,

ITEK, McDonnell, and many others including Radiation/Harris as a subcontractor to Perkin-Elmer. The main camera system involved large optics and sophisticated film handling mechanisms. The film was state-of-the-art from Kodak and was on two 8 foot diameter supply rolls, and over 25 miles long for stereoscopic viewing. By the way, the film image frame was 6 inches wide so you can imagine the issues of manipulating this film to move it past the focal plane at 200 inches per second in proper synchronization on two cameras one facing forward and one facing aft. That is what Radiation's controls and servos did. After exposure this film was wound into canisters in one of the four re-entry vehicles. The CIA was the driving force for our portion of this program with heavy Air Force participation on the balance. The prime contractor for Radiation was the Perkin-Elmer Corporation as the provider of the primary optical system.

"We provided virtually all of the on-board electronics in 14 Work Areas, as they were called. These ranged from the processing and control to the servo signals to the film transport motors, plus several other electronic functions. The electronics were packaged in welded cord wood planar mounted modules Radiation had developed for aerospace products.

"While we won this job on the basis of our packaging design, we soon realized that the Perkin-Elmer engineers were out of their depth in electronic system and circuit design so we proposed that we take on those roles. We had the support of the CIA, plus firm direction from Harold O'Kelly in this effort, and we were successful in greatly expanding our scope to include almost all of the electrical design. When we told Harold of our concerns in this area, he said 'GO GET IT ALL.' So we did. To do this, we invaded PE with our engineers, and after significant resistance from the affected parties there, we took over the circuit and electronic system design responsibility. It was tough under the security constraints, but our team on-site transferred the needs of the optical systems to our engineers in Palm Bay with sufficient knowledge to make the system work reliably. Prior to this time, we were being blamed for schedule delays that were actually the result of their electrical engineer's immature and inadequate designs constantly being changed thus negating the progress on our packaging designs. We solved this when we took over the all-inclusive role by forcing the signing of a baseline document known affectionately as BOO-2, wherein all changes from PE would be a change in scope to the contract.

"(The identifier BOO-2 came about as a result of my scribbled 1300-2 which was misread by the secretary because my 1 and 3 were too close together and was published as a B instead of the two numbers 1 and 3. It quickly showed up in too many places so it stuck.)

"From the time of our first proposal in the summer of 1966 and first $60K study contract that fall, the scope of the program grew to about $13M at the end of 1967. We also had trouble staffing the program as two other major programs (W-71 and MOL) were very active in the Aerospace Division, fighting over resources. The Harris/Radiation merger had taken place then so it was chaotic times to say the least, but our team grew and stayed focused on the schedule with some very long hours and interesting events that would lead to too many war stories. The 1300 Program grew to the point that it merited a director and A. J. Scott was named to that position. I have been told that the total orders secured as a result of this program were over $100M.

"The total program was under extreme pressure and there were many visitors, trips and reviews by those concerned. We made frequent use of the Lear Jet for weekly trips and had many customer and other reviews with what we now know was the CIA. Those few in the know had to convince the rest of the internal organization that this program was real and very important, as this was the era of Herb Schiller, this was not easy, but Harold believed in us, and soon everyone else did.

"The gung-ho spirit of Radiation came through as we evolved into Harris and we got the job done such that the total program was very successful from the first launch in 1971. The capability for space reconnaissance for wide coverage and high resolution was realized from the start by the national intelligence leaders. For example, each pair of frames yielded a stereoscopic view of a swath of 370 nautical miles—from Cincinnati to Washington DC—and a resolution of 2 to 3 feet—if you looked at an army deployment, you could count men. According to documents released by the NRO, each HEXAGON satellite mission lasted an average of 124 days, with the satellite returning four film capsules in sequence to send its photos back to Earth. An aircraft would catch the return capsule in mid-air by snagging its parachute following the canister's re-entry. The NRO launched 20 KH-9 HEXAGON satellites from California's Vandenberg AFB from June 1971 to April 1986, and all missions were successful except one when the last rocket blew up on the pad.

"The KH-9 was wildly successful throughout the seventies and into the eighties. With only a few glitches along the way, KH-9 satellites operated outstandingly well right from the start, despite the fact that according to everybody who worked on it, it was—and remains—the most complicated mechanical device ever launched into space. The Swiss make finely engineered watches, and the Americans build finely engineered reconnaissance satellites. As one of the major programs in the transition from Radiation to Harris we can be proud of the role we played in the defense of our nation and a peaceful end to the nuclear war threats by the major powers.

"Thanks to Jack Pruitt and Richard Burner for their help replacing 45 year old memory lapses and editing this summary."

The above material was presentation at the Radiation Old Timers Gathering April 2012 by Scott Broadway.

Dave Claycomb provided several picture of the 1300 units in final assembly. The technology was based on "cord wood" modules, in which discrete resistors, transistors and other components were connected by welding metal ribbons to their leads. The modules were all a standard size (approx. 3.5 x 2.5 x 1.5 inches) and once the leads were all welded together the module was tested. Once passing the initial test, the module was placed into a foaming fixture, foam mixture added and placed in the oven to cure. The completed modules were then mounted/soldered onto the printed wiring boards, the printed wiring boards were wired together, the unit was tested, and compression loaded under temperature. The final assembly was then acceptance tested and shipped.

It sounds complicated, and it was, but it resulted in a compact, rugged, and reliable assembly.
Program contributors: Chris Catsimanes, Scott Broadway, Dave Claycomb, Jack Pruitt, Richard Burner.

Igloo White

One of the important programs at Radiation during the Vietnam War was Igloo White. It was a covert electronic warfare system aimed at detecting vehicle and foot traffic along the Ho Chi Minh Trail, which delivered war supplies to the Viet Cong in South Vietnam. It was used from 1968 to 1973.

Igloo White consisted of three parts: a network of sensors; an airborne relay station; and a ground data processing center.

Several types of seismic and acoustic sensors were used. Most were dropped from aircraft, but there were also helicopter-dropped and manually placed sensors. The seismic sensors were spike-shaped and designed to penetrate the ground, leaving only the antenna exposed. Some of the acoustic sensors were dropped by parachute and were intended to operate while suspended in trees. Other acoustic sensors were combined with seismic sensors on the spike-shaped versions.

The relay aircraft were originally four engine EC-121's, which carried a large crew who performed preliminary assessment before transmitting selected data to the ground. Later in the war, the expensive and vulnerable EC121's were replaced with modified Beech A-36 Bonanzas, called QU-22B's, which did not perform on-board processing, but relayed all the data to the ground. Originally these were intended to be remotely piloted, but always carried a pilot on operational missions.

The ground station (sometimes called "Music Box") was a large, tightly secured building located on Korat Royal Thai Air Force Base at Nakhon Phanom (NKP). Here banks of computers and displays were used analyze the sensor data to detect supply convoys and estimate their size and movement. The results were used to call in various kinds of strikes on the supply trains. Security at the ground station was so tight that janitors were not allowed inside and the analysts had to perform this additional function.

The system experienced a number of problems. Not all sensors were inserted where planned, and not all survived the rigor of emplacement. It turned out that a bird common to the region, the Nightjar, had a call that spoofed the sync signal of the up-link, causing mysterious system failures. Often the complex system worked as planned and tales were told of hearing truck convoys start up and proceed down the trail, followed by the sound of them being destroyed by bombs. All this was played back to impress the brass in Washington. Another recording is claimed to document a conversation between Viet Cong soldiers attempting to recover a sensor parachute as dress material for a girl friend.

Radiation was responsible for the down links between the orbiting EC-121 relay aircraft and Music Box, and had full integration responsibility for the QU-22B relay aircraft under a program called Pave Eagle – both necessitating the deployment of a large field support crew to NKP. Radiation designed and built large numbers of cylindrical sensor elements, referred to as "common modules", presumably because they were used in a variety of sensor types. A separate operation was set up in Miami for manufacturing these common modules.

A follow-on program was DART (Deployable Automatic Relay Terminal.) According to A B Amis, "DART was a quick reaction program (90 days from contract signing to delivery, as I recall) for a "Deployable Automatic Relay Terminal" that had an S-band receiving/tracking antenna mounted up on a tower, and a trailer full of electronics ending up with marks made on a paper tape recorder about the width of newsprint, from which trained observers could hopefully recognize patterns of sensor firings indicating VC movements along the Ho Chi Minh trail. Sort of a miniature version of the NKP Music Box installation, less the IBM computers. Tom Sheffield was responsible for the paper tape recorder, and Frank Perkins was responsible for a bit rate synchronizer [which operated at the then blazing speed of 1.8 Mbits per second.] I spent a couple of weeks up at Duke Field at Eglin with that system during sell-off to Mitre, "

Space Deployable Antennas

Various defense systems require large parabolic antenna on spacecraft in orbit. The antenna are much too large to fit within the fairing of the launch vehicle, so they must be folded up like an umbrella for launch and opened in zero gravity in orbit. Most applications remain shrouded in secrecy after 50 years. Bill Corry was one of the major design contributors to the programs, and he recently wrote the following essay.

Remembering Deployable Antenna Technology
W. A. Corry

In the early 1970s, Radiation (with Gene Cross the Program Manager) was asked by a military sponsor to "do something with" a foldable 4-meter rigid antenna that was being offered by a company I think was named Neotech. The several solid metal petals folded from outside to the center to create a stack of petals that were to unfold like a flower to create a solid surface reflector for space. When the device failed to work properly, we were asked to look for a solution using multiple rigid ribs that folded and unfolded like an umbrella with an elastic metal mesh surface stretched between the ribs.

Our first efforts used 24 curved ribs rotated from a central hub using a central ballscrew mechanism developed by Dwight Gilger with separate linkages for each rib. Early designs used a Dielguide antenna feed and support structure to which the outer ends of the ribs were attached when the structure was stowed for launch. Even then, the resonant frequency of the resulting structure was much too low. The design was subsequently upgraded with larger ribs that crowded the center deployment mechanism and increased the overall stowed diameter. Still, the structural resonant frequency was too low to survive a rocket launch. More support was needed.

By adding an additional rib support half way up the stowed structure, Carl Anker demonstrated that the resonant frequency problem was solved but with another level of mechanism complexity.

An additional study was proposed to the customer to improve the accuracy of the surface. Adding more ribs to support the stretched metal mesh was one concept. The winning concept involved the use of a mesh surface behind the front surface and attaching the front and back together using a series of ties to form the front surface into the desired (parabolic) shape. This was a game changer in that the development of surface accuracy was no longer a question of the number of ribs, but rather a question of how many ties are used between the front and rear mesh surfaces. With enough ties between the two, surface accuracies approaching solid surfaces could be achieved.

The next big challenge was the development of the metallic mesh surface to be elastic enough and reflective enough to work with the feed system. This problem became a major obstacle over many years to develop a gold plated molybdenum wire mesh that could be knit with commercial machinery. During development of the elastic metal mesh, there was some serious questioning of our methods when we purchased several pairs of panty house to simulate the surface.

Early RF measurements revealed an even greater and unpredictable issue with the surface of the back mesh interacting by reflecting the signal that escaped through the front mesh and causing severe problems with the wavefront. Various techniques were considered to minimize this effect by reducing the size of the mesh on the back or making it non-reflective. Neither worked. The answer, suggested by a customer was to replace the rear mesh with a series of strings attached to the back of the ribs from which the ties to the front mesh could be attached. We adopted the name of "strings and things" to describe this concept which has been used on all subsequent antenna designs.

Having resolved most of the structural issues, the next hurdle was in the details. Aluminum ribs, even when wrapped with multi-layer insulation (MLI), lacked the thermal stability necessary to keep the surface focused when exposed to solar radiation and the cold of outer space. A major development by Hercules Aerospace guided by Marvin Sullivan achieved a remarkable solution with Graphite Fiber Reinforced Plastic (GFRP) ribs due to their very low modulus of thermal expansion. Even so, these needed to be wrapped with MLI to achieve the requisite thermal stability.

Special tooling to compensate for earth gravity was necessary to set the surface so that it assumed the correct shape after withstanding a severe launch environment and deployed in zero gravity with full solar irradiation. The resulting non-linear analysis led by Dr. Jim Sturgis, key in determining what length to set the hundreds of individual ties between the rear strings and the mesh surface, was ground-breaking technology.

Hexagon Satellite on display at Smithsonian

Radiation-built hardware from Hexagon, before final assembly and compression.

"Common Modules" for Igloo White.

Sensors for Igloo White

Somewhere during the above development, I replaced Gene Cross as Program Manager for the Deployable Antenna Technology Programs with Dr. Bill Tankersley my close associate and best friend as Systems Engineer. It seemed there was a classified application for this technology, that resulted in the hiring of Dr. Jim Sturgis, and Dr. Allen Henry, who, under the guidance of Giles Sinkewicz won and successfully executed a substantial classified program about 1974. I seem to remember my brother, Doug being involved as the Program Manager while I gave up my Program Manager status to Bill McCaslin to return to my first love, structural mechanisms design. Enter Russ Ford, Designer Extraordinaire, who could translate and embellish our design engineer's work into manufacturable hardware. We worked together for 20 years. He made me look good.

Success of the Deployable Antenna Programs was in the details. New concepts to support the structures during launch and releasing them to deploy without snagging the mesh were achieved by a cadre of engineers and analysts of a high caliber. Oh, yes, the antenna had an RF function as well, certainly not an incidental. A partial list of contributors included Al Henry, Bill McCaslin, Bill Tankersley, Jim Sturgis, Bill Thrasher, Carl Anker, Dave Burt, Dwight Gilger, Eddie Faust, Gene Cross, Homer Bartlett, Giles Sinkewicz, Russ Ford, Tom Tonk, Gus Gossard, Harvey Ulmer, Gene Wooten, Doug Dilts, Jim Erikson, Jim Sturgis, Marty Schwam, Marve Sullivan, Rick Harless, Paul Dixon, and Tom Mills.

The NASA Technical Data Relay Satellite needed two 5 meter deployable antennas to communicate between the Shuttles and Ground Stations. Harris built thirteen of these, two of which were lost when the Shuttle Challenger went down. Harris also built the TDRSS Ground Stations.

Over 40 years have elapsed since my efforts in designing Deployable Antennas. The memory has suffered since. Please forgive my lapses and additions to what actually occurred.

The successful design and fabrication of Deployable Antennas goes on today.

Stonehouse/Bayhouse

By way of background, I quote from *internet postings by Sven Grahn, a Swedish space historian.*

"In the early sixties the United States created a sophisticated system of ground stations to monitor the flights in the Soviet lunar and planetary exploration program. This was a joint effort between NSA, Norad, DIA's Defense Special Missile and Astronautics Center, CIA's Office of Elint and Foreign Missile and Space Analysis Center.

"A network of stations was set up, but the main station was placed at Asmara - in then Ethiopia - at almost the same longitude as the main Soviet ground station in the Crimea. It had an 85 ft antenna (operational in April 1965) and a 150 ft dish with lower surface accuracy (operational in 1964). A search on the WWW reveals that there was a US Army Security Agency station at Asmara, nicknamed "Kagnew Station", starting in May 1950. Presumably the deep space monitoring station was co-located with the army facility.

"The first spacecraft that the Asmara station tracked was Luna 5 launched in the direction of the moon in May 1965. Asmara picked up "both of the two spacecraft signals", measured Doppler shift and could not find any evidence of retrorocket firing. Asmara also tracked Luna 6, but missed Zond 3 because it did not obtain a "fix" during injection, with which to determine the trajectory further away from the earth."

Grahn also cites a Department of Defense news release in January, 1964 which gave the cover story for Asmara--a new deep-space research site--Stonehouse:

"Experimentation in the peaceful uses of space will receive added impetus in Africa with the installation, at Kagnew Station, of additional equipment for space communications research and for future study of radio receiving and transmitting techniques. The new equipment, now ready for installation, will consist of two rotatable parabolic antennas, one 85 feet in diameter and the other 150 feet in diameter. These modern antennae are designed to further the study of long-range communications and to facilitate the study of the effects of the ionosphere on communications. The selection of Asmara for this important new space research activity resulted from extensive surveys to find an area combining relatively quiet electronic environment, and suitable topographic features and climate characteristics, near the equatorial belt. Kagnew Station is a particularly appropriate site to receive the new antennas in light of the stations past contributory research into natural electronic phenomena. The new equipment will expand Kagnew's communications research capability and will permit scientific measurement of unusual transmission characteristics in outer space communications research. United States interest in this research activity is based on the desire to improve long-range communications world-wide. The new installation will make an important contribution to man's expanding knowledge of the mysteries of outer space. Materials for the new antennas will begin to arrive at the seaport of Massawa in early May. From there, they will truck-hauled to Asmara. The installation is expected to be completed in 1965 and during phases of its construction should employ many Ethiopian workers. Arrangements will be made for groups of visitors to tour the new facility during its construction in accordance with past practice at other parts of Kagnew Station."

Radiation's Role

The name "Stonehouse" was originally applied by the government to the the "Deep Space Research Site" at Asmara, a joint project of various US military and intelligence organizations. The station operated from 1965 to 1975. Within Radiation, "Stonehouse" was applied to the 85 foot antenna and "Bayhouse" to the 150 foot antenna.

At the time of award, these two programs represented the largest, most complex project ever undertaken by Radiation's RF Department. Both contracts were awarded in 1963, with Bayhouse following Stonehouse by about 30 days. Both of these systems were intended to carry out an important SigInt mission for the US Government. Even after nearly fifty years, there is much about this project which is still classified. What follows is an unclassified description of both

Stonehouse

The Stonehouse system as an engineer's dream come true. Every element was at or near the state of the art at that time, and it included virtually everything that a signal intelligence system designer could think of. The antenna was an 85-foot diameter dish mounted on an X-Y pedestal. It had seven interchangeable feeds, in both apex-mounted and cassegrainian configurations. A feed focus mechanism provided three feet of travel and positioning within .010 inches accuracy. Low-noise amplification was provided by a cryogenically-cooled maser for each frequency band. Precision antenna pointing control was achieved by torque-biased hydraulic drives on each axis. Shaft encoding was provided by 21-bit Grey code encoders on each axis.

Other features on Stonehouse included a Bausch & Lomb boresight telescope and camera, electric auxiliary drives, and a two-stage vacuum pumping system to support the maser' cryogenic containment vessel. RF alignment was achieved using 200 foot near field and 400 foot far field boresight towers.

Bayhouse

The Bayhouse system was not quite so technically sophisticated, but nevertheless represented significant challenges. The Antenna was 150 feet in diameter, much larger than anything Radiation had ever built. Its pedestal was a "wheel and track" structure, providing Az-El coordinates. This antenna was a modified version of a radio telescope originally developed at Stanford University. Bill Corry redesigned the Elevation axis drive system to provide a geared wheel segment instead of Stanford's linear actuators. The lessons learned from Bayhouse enabled the development of the Altair 150 foot antenna three years later.

The entire Bayhouse structure rolled around on a 110 foot diameter railroad track. The structure was large enough that all of the system electronics could be housed in a modular building on its rotating base.

Deployment

Radiation was responsible for the installation of the antennas at the remote site. This invovlved planning and procuing all supplies and equipment, packing it in a ship and trucking it up the mountain to the site.

Stonehouse and Bayhouse were installed just outside Asmara, Ethiopia (now Eritrea) in the summer and fall of 1964. The site was delivered and became operational in the spring of 1965. Crews from Radiation operated and maintained the systems for another decade. When the regime of Haile Selassie ended and his son became the ruler, the site was no longer considered safe for expatriates. The site was closed down, the people evacuated, and the sensitive equipment was removed. The remaining hardware was scuttled and scrapped. The site now shows no evidence of its prior use.

The combination of secrecy, high priority, technical complexity and remoteness generated a spirit of adventure and deering-do in the Radiation staff, as illustrated by the many stories of the program, under "Stonehouse Tales" in the appendix.

Standard Products

A recurring dream of Radiation (and probably other systems houses) was to cash in on some unique product. The idea was that some customer would hire us for a system including some widely useful product. With the development expenses paid, we could then profitably sell vast numbers of the product from manufacturing while our design engineers continued to work on custom systems. This seldom worked out, mainly because of the drastic differences in marketing product and systems. However, a few product-like devices are of interest because of what they were and where they led. There were also a contractual advantage to having a product cataloged with a list price.

Radicon and Radiplex

My initial assignment at Radiation was to shepherd this pair of products. Radiation had built a number of data acquisition systems and was a leader in the multiplexing and A/D conversion technology, so it seemed a natural. Radiation had hired a prominent Industrial Designer to design a sexy package and funded a program to build a dozen units. When I arrived on the scene, the funding was running low and the program was in trouble.

One problem was that the Radicons had suffered uncontrolled "feature growth" common in units designed by committee. In an effort to make it "universal" the Radicon had multiple input ranges, operating modes and output formats. As a result, complexity had grown and manufacturing and test had become time-consuming and expensive. Internally, Project Engineers had revolted against the bloated design and insisted on versions that only incorporated the features need for their systems. Further, technology had outrun the design since it was frozen a year previously and higher performance was available with cooler-running, more reliable circuitry. Most internal users had opted for these improvements. None of this was immediately obvious to me, and I was at somewhat of a loss as to what to do. Fortunately, I decided to talk to the project engineers who were using the units. This had a number of good results. I met a number of interesting, skilled engineers who became good friends, and learned a lot about how the company operated. I also learned to troubleshoot and repair the power supplies in the Radicon. Most of them oscillated at high level, which upset the accuracy and operation of the unit. The fix was a change in a capacitor value in the regulators. I don't know how this error got into the design, but it was useful to learn how to fix it.

I don't think we ever sold a fully featured unit, but I was able to define a more practical unit with a useful subset of characteristics which was useful internally.

The Radiplex standard product suffered from technology advances during its development. It was designed as a high level multiplexer, that is it handled signals of a few volts, because the diode switches used did not have sufficient accuracy for lower levels. Most transducers, such as thermocouples and strain gauges, generated signals in the millivolt range, and it was highly desirable to multiplex them before amplification. Alloy junction transistors had come on the market whose characteristic were suitable for low level multiplexing, and a low level multiplexer (the Radiplex 89) was developed for the Thiokol data acquisition systems.

Radiation-at-Stanford, our California Division won a job for a compact data acquisition cart they called Radatac, taking advantage of the Radiplex 89. Unfortunately, there was confusion in some of the detailed specifications of the multiplexer and they had promised some features which the Radiplex could not deliver. Also, their Radicon had some of its usual teething problems. I ended up making several emergency trips to the west coast to try to get things working in an acceptable manner. In those days our only access to transcontinental jet service was to fly propeller flights to Chicago. It was worth the detour to have the fast, smooth jet service from there, particularly when I discovered that no one would complain that I upgraded to First Class for the Chicago-San Francisco leg.

Radatac was in a big lab where Levinthal was working on some high power equipment. I was working on the coder when there was a loud Klaxon blast, followed the louder crack of a man-made lightning bolt. It was Levinthal testing discharge circuits for a high voltage capacitor. The horn was intended to warn you of the impending bolt, but was almost as startling as the bolt itself.

Unfortunately, Radatac also suffered from design problems external to the coder and multiplexer. The last gasp attempts to make it work were the night before it was to be demonstrated at an trade show in Los Angeles Eventually it was shipped back to Melbourne and Bernard France redesigned it completely. We lost a ton of money, but the customer was eventually satisfied.

Not long after, Radiation-at-Sanford was sold.

Logic Modules

On of the more successful Radiation products were logic modules, small encapsulated, plug-in assemblies. The conventional packaging consisted of plug-in printed circuit cards each containing a few gates or flip flops. Users maintained their systems by replacing faulty cards from a stock of spares. These were either repaired in house (which required dedicated technicians, test equipment and replacement components) or new cards were purchased from the manufacturer. Modules originally contained only one expensive element, a transistor, which was also the most likely element to fail. A failed module could be simply discarded.

It was amazing what an emotional reaction was attached to the choice between cards and modules, both internally and externally. Some customers were won over by the logistic arguments, or just by the novelty. A few people in Radiation management were adamantly opposed, for reasons that were never clear to me. At one point there was a threat to deny approval to ever build another system with modules. A call to a good customer with a pending module system contract soon put an end to that threat.

Module design went through a series of evolutions. In the original design, solder-assembled circuitry was potted in a round epoxy module that plugged into a miniature tube socket. The tube sockets were mounted on a large vertical panel, which were slide mounted in the equipment racks. Back panel wiring was hand routed and soldered.

Some Radiation Standard Prooducts

Rectangular Modules

Radicorder fixed styli recorder

Radicon A/D converter

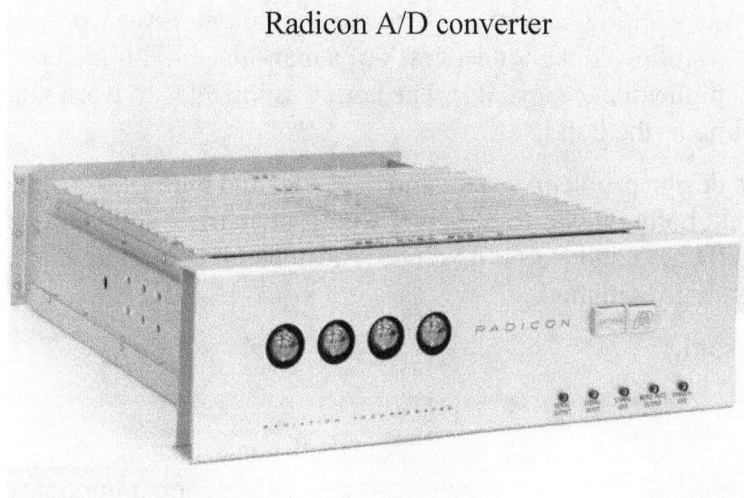

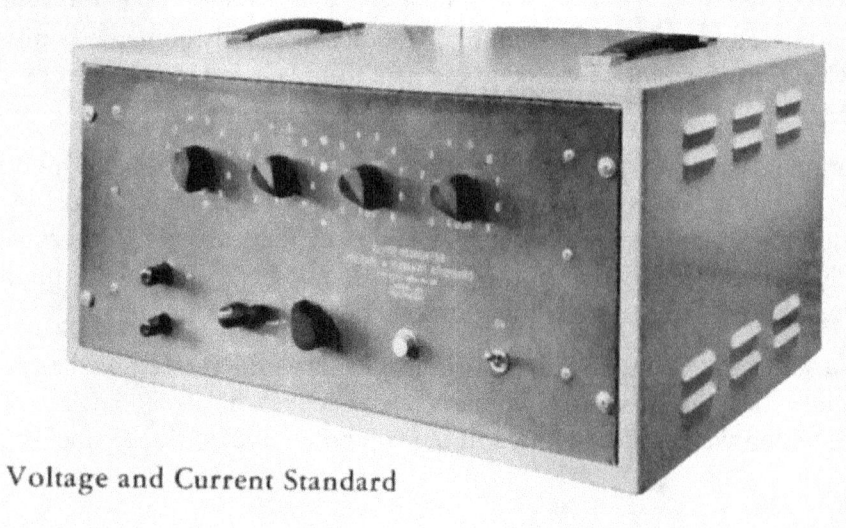

Model 3115 Transmitter

MDC12A voltage and current standard

Voltage and Current Standard

A considerable number of satisfactory systems were delivered in this design, notable to Bell Labs for the Nike system. Improvements were made in both the module fabrication and the panel mounting. Welding was adopted to assemble the components; this was more reliable and required less skill from the assembler. Efforts were made to use conductive epoxy for the mechanical and electrical connections. This method resulted in modules that would fail after some use, but would then heal themselves after a period in a "hospital" box. This was clearly unsatisfactory and conductive epoxy was abandoned. The welded modules were assembled on rectangular headers, with blade-type connectors. These plugged in to panels with wirewrap connectors. Wirewrap connections are made with a special tool which wraps a section of solid wire (with a length of insulation removed) around rectangular posts with sharp corners. The force of the wrapping action almost welds the wire and post together, forming a rugged reliable connection. Wires can be replaced by unwrapping them with a special tool. Panels can be wire wrapped with a special tool which automatically reads a wire list, cuts, strips and wraps the wires without intervention (or errors!).

When integrated circuits became available, a miniature version of the module was created, containing a flat pack, but when dual in line (DIP) packages became available, they could be directly plugged in to a wire wrap panel, and welded logic modules finally became obsolete.

Radicorder

In about 1956, Orlando Instrumentation Division received contracts from Wright-Patterson AFB for graphic recorders, one flight recorder and one lab recorder. At the time graphic recorders were cumbersome and inconvenient. One type used meter movements linked to ink pens. These were limited in frequency response and messy and unreliable. Photo galvanometer type recorders light beams deflected by mirrors to write on photographic film. These were higher in performance but still had moving parts and the film had to be developed before it could be viewed. The Air Force recorders used an array of fixed styli to write on elector-sensitive "Teledeltos" paper and had no moving parts. The contrast was not great, but the accuracy was good and the plots were available immediately. A rack mountable version of this recorder was named the Radicorder and used in a number of ground data systems. In at least some systems alphanumeric legends could be printed directly on the plots.

This technology was later applied to make a very high speed computer printer, described elsewhere.

MDC12A Current and Voltage Standard

Another standard product that came out of Orlando was MDC12A Current and Voltage Standard. I remember using one a few times in the lab, but don't have any information on external sales.

Model 3115 Transmitter

This was a small FM telemetry transmitter. According to A B Amis it had a good reputation for ruggedness and reliability, and around a hundred were sold.

Model 540 PCM Decom

This system was sold as a standard product so that the customer could get exactly what he wanted without an elaborate competition. It was used to test the various stages and vehicles for the Apollo moon landing program. It was installed at the North American Aviation plant in Downey, CA, at Kennedy Space Center and a GE plant in Daytona Beach.

Chapter 6
Other Programs

The main common thread of these programs is their technical; or historical interest. Some were significant to the company, others turned out to be less significant, but all deserve to be remembered.

Doppler Navigator

Harold O'Kelley writes: "My first assignment was on Project 1130, "Doppler Navigation Radar for Army Helicopters." I acted as a systems analyst evaluating the spectrum of the signal back-scattered from the ground and returned to helicopter radar. This was because my bosses, because of my prior professorship, looked upon me as a scientist rather than engineer. Even though the prototype equipment was tested in a fixed wing aircraft, it never was placed in production—the prototype did not work very well. Actually, the project ended ignominiously when the test aircraft made a bad landing, severely damaging the radar and slightly injuring Warren Wiener, our Project Engineer, leading the effort."

A.B. Amis writes: "John Downs had been hired from Sperry Gyroscope and made head of RF Division, probably largely on the strength of his having brought with him a Doppler Navigation Radar project with Sperry as the customer (Project 1130). Progress seemed to be lagging on that project, which had been staffed six months or a year earlier primarily with new-graduate engineers (Jay Fleming was one of them), so Project Engineer Warren Wiener had gone to his management with a request for some more senior people. Ed Dorsett, Walt Minnerly, and I were assigned to the project in late 1958, and perhaps Guy Pelchat also – although Guy may have already been working on the project – he'd worked on a similar project at Canadian Marconi before coming to Radiation. I'm fuzzy on this, but we may have started off designing the Doppler navigator around micro-miniature vacuum tubes, but transistors were just coming on the scene at that time, so the whole crew attended lectures given by Bill Eddins on transistor circuit design. I took on the task of trying to design a transistorized 30 mc IF strip, and I guess I must have come up with something sort of workable because the project progressed on to flight testing some time later -- but the flight tests ended abruptly when the landing gear on the test aircraft collapsed while the plane was taxiing, smashing much of our equipment which protruded from the bottom of the aircraft in a radome. Project 1130 was initially housed in Bldg. 9 at the airport when I joined it, but moved shortly thereafter to the Malabar test facility. I actually remember far more about the people that I worked with than about the technical details of that project. Ed Dorsett and I had already been singing together in a barbershop quartet, and after assignment to work together on 1130 we managed to corral a couple of other co-workers (Dick Ferguson and Bob Fletcher) to "harmonize" with us during lunch breaks or other times when the four of us found ourselves together with a little time on our hands."

Tornado Detection Radar.

The Tornado Detection Radar, predecessor to today's Doppler Weather Radar, was built and tested at Malabar (a small tornado showed up right on schedule -- ask Jay Fleming).

Harold O'Kelley writes: " Next John Downs assigned me to a project for the U.S. Weather Bureau, a Doppler Radar. Doppler radar displays are seen when we watch our local weathercast.

"The term "Doppler" (a man's name) means frequency shift caused by moving objects. County Mounties use Doppler when they catch you speeding in your automobile. When I joined Radiation, Doppler radar was not widely used. Then the military used it for measuring aircraft ground speed.

"The Weather Bureau theorized that Doppler radar might detect tornadoes. A tornado touching the ground picks up trash and twists debris with speeds guessed to be 200-300 miles per hour. Doppler Radar only detects a frequency shift from objects having a velocity component directly toward or away from the radar. Reflections from objects moving perpendicular to the radar-beam do not have a frequency shift. Therefore, a funnel cloud filled with swirling debris would show velocities from zero to, say, 300 mph. The Weather Bureau wanted to test their theory.

Doppler Navigator installed in aircraft for flight testing.

Test aircraft after accident, which ended the program.

"Further, they hoped to record the velocity spectrum of a funnel cloud to determine actual speeds. I was the Project Engineer at Radiation designated to design and build the Weather Bureau Doppler radar. We took an old radar trailer, supplied by the government, and used it to house the electronics. We put two new dishes on the antenna mount, and made it into a X-Band CW Doppler Radar. (X-Band means that it operated at about 10,000 megacycles—megahertz—and CW meant that it sent out a continuous wave instead of pulses.) The acceptance test was to show that a BB shot directly at the radar from a distance of 100 yards was detectable. My radar passed easily.

"From this project I earned the nickname "Overrun O'Kelley." Marketer Myron Roebuck, who later became Vice President of Marketing, called me that. The fixed price was $30,000, and cost was estimated at $27,000. I brought in the finished product for $28,000. I still respond to the name when Buck calls me.

The radar was used by the Weather Bureau for tornado studies in the southern plains during the late 1950s. It was turned over to the National Severe Storms Laboratory in Norman, OK, where it was converted to a pulsed Doppler radar in 1964 to allow range measurements which the CW radar could not supply, and eventually lead to todays weather radars.

Doppler radar at Malabar

Bulllpup Ground Pilot Trainer

This system was designed in Orlando to train pilots in the operation of the Bullpup missile. The Bullpup was the first mass-produced air-surface command guided missile, first deployed by the United States Navy in 1959 as the ASM-N-7, until it was redesignated the AGM-12B in 1962.

The Bullpup had a Manual Command Line Of Sight guidance system with roll-stabilization. In flight the pilot or weapons operator tracked the Bullpup by watching a flare on the back of the missile and used a control joystick to steer it toward the target using radio signals. The Trainer consisted of a 17 inch CRT on which was displayed a simulation of the missile tracking flare. Tracking commands generated by the trainee corrected for error generated by the Trainer.

An early video game!

TSC-54 (Mod V)

The AN/TSC-54 was a satellite terminal designed for quick-reaction use with the military SATCOM system. It was transportable in two C-130 cargo aircraft, and required only a few hours for assembly and the initiation of operations. The initial contract for Mark V's was awarded in 1965.

I can remember huddling in a Mark V trailer in the back lot at Palm Bay, running bit error rate tests with a breadboard error correction coder/decoder--some of the earliest such measurements on a live satellite link (in1958)

Quick Reaction Programs

A repeated theme that occurs in peoples memories is a job which, due to some customer emergency had to be delivered in a short time frame.

Thiokol System

The earliest one I remember involved a Thiokol data acquisition systems. One of the minuteman motors being tested blew up, damaging the test cell and destroying our data acquisition system. Bits and pieces of systems for future delivery were available, and a replacement system was delivered in a week or two, with lots of long hours.

AMARS

Carl Swisher writes:

Speaking of projects with short fuses, I worked on several green tag jobs [jobs with relaxed inspection and quality control requirement] with short fuses but the one with the shortest fuse was the AMARS project. Here is a brief History of AMARS.

The AMARS Project. (Automatic Multiple Address Routing System)

"USA STRATCOM operated several torn tape relay stations around the world handling messages in teletype paper tape format. The relay station was a collection point for paper tape messages from outposts and a distribution point for sending paper tape messages to other connected outposts. (This was in the days before computers.)

"When a paper tape message came into the relay station it was walked across the room and loaded into the paper tape reader that the message was routed to. The reader then read the message and it was punched out at the remote receiving outpost. It was a very simple process except when the incoming paper tape contained multiple routing instructions. (For two or more outposts.) The operator at the tape relay then had to make multiple copies of the message with each copy contained only the routing instructions for each outpost. Because of the high volume of multiple address messages being processed at the torn tape relay stations in Vietnam the relay stations were severely backlogged and critical messages were not being retransmitted.

"The STRATCOM Engineering office contracted with some company in the DC area to build a MAPU, a Multiple Address Processing Unit. This company was not able to deliver a working unit and the contract was canceled. Somehow our intrepid B & B marketeer Mel Skinner found out about this situation and assured STRATCOM that Rad Inc could perform. STRATCOM agreed to award a sole source contract to Rad Inc if they would agree to a 45 day delivery with heavy penalties for non-performance. Al Johnson and Julian Scott asked me to review the SOW and asked me if there was any possible way to deliver in 45 days. In the meantime Mel had suggested hiring Stan Gombos, who had worked on the failed MAPU project and who at least understood the problem. After discussing it with Stan and making a trip to Seattle to talk to Tally Corporation (a supplier of high speed paper tape readers and punches) I stuck my neck out and told Al that there were two chances we could deliver on time and make it work. The two chances were slim and none with the slim chance being if the project were giving top priority by management, which they agreed to.

"Thus was created the AMARS project. Dan Mestayer, Stan Gombos, myself, technicians Cecil Martin and Billy Joe Sutphin and several wire girls started on a 45 day, around the clock work schedule. We only went home to take a quick nap and change clothes. Our wives brought food to us and left it at the gate.

"There were no weekends, 45 straight working days. Several planners, including Jim Pickett, Jim Maxner and Harry Goode begged, borrowed and stole logic modules, trays, racks, etc. The shops worked overtime creating the special modules we needed to program the routing instructions. Jeannette Fanus wired wrapped the trays for the routing instructions with zero errors and was given a award for her performance. Tally Corporation delivered the high speed tape punches (30 each) and readers (2 each) on time.

We started checkout with only hours left before our scheduled acceptance test with the customer. When we fired up our scope and pushed the start button and put the scope probe on the first modules nothing worked as designed. Indicator lights started flashing, punches started punching when they were not supposed to, it looked like the beginning of a disaster. A quick review of the design on paper did not show any design errors so what was going on? It turned out that a simple rule had been forgotten, "never leave an unused input floating." Cecil picked up his wire wrap tool and stated tying down the unused inputs and things started clicking. Hours later checkout was complete and we all went home to sleep and wait for the customer the next morning.

"Acceptance Test with the customer the next morning was completed before noon on the 45th day and the 13 racks of equipment was packed up by Bob Eisinbenger and taken to Patrick to be air shipped to Saigon. Now came the big question. Who was going to get on an airplane and head to Saigon to install the equipment and put it in operation? Since the Marines always go in first, guess who was elected?

"I was on an Airplane to Saigon two days later and it took 45 days to get the equipment installed, programmed and checked out. The Strategic facility was outside of Saigon near the village of Phu Lam. The year was 1966. The CO of the facility was Lt Col SS Ashton Jr, a West Point graduate. We became close friends and have visited and maintained this friendship ever since. In May of 2011 my wife and I made a trip to West Palm Beach to visit Col. Ashton and his wife. He will be at the 60th reunion of his West Point class in June 2011.

"The Phu Lam Amars was the first in a series. Other Amars were installed later at the Air Force Com Center at Ton San Nhut air base in Saigon, and Stratcom facilities in Okinawa, Nha Trang and Da Nang, VN, Bangkok and Korat, Thailand and a Stratcom facility in Germany.

"I was at a party in Cape Canaveral just a few short years ago and was talking to a fellow who said he was a retired Army S/Sgt. I asked him where he served and what he did. He told me that I would not know where it was that he served or what he did if he told me but I said, go ahead and tell me. He said he was an AMARS technician at Phu Lam. Small world isn't it?"
Carl

Other quick reaction jobs occurred after the Harris merger, but they reflect the attitude and approach from Radiation days, so I include them.

The 10 Day Wonder
Wally Norris
The earliest one I remember was when, on November 4, 1974, Harris received the go-ahead to provide a replacement for the subcarrier demodulator in the UK/TCS RAF Oakhanger Site in England. The station had been built by Harris as a "rack and stack" job—all components simply purchased and integrated. Systems engineering on such a job consists of looking through vendor catalogs. The station was to support the Skynet Communications Satellite scheduled for launch on November 19, and repeated tests showed that the subcarrier demodulator would not hold lock at the expected received power level. This downlink was similar to a Space Ground Link System (SGLS) Carrier 1 telemetry and ranging signal. The modulation of this signal is often called PM/PSK. The return pseudo-noise (PN) ranging code is direct phase modulated onto the S-Band carrier. Telemetry and sometimes communications are PSK modulated onto one or more subcarriers, which in turn phase modulate the main carrier. I don't think Skynet used the ranging code, and the only subcarrier was 1024 kHz with 250 bps PSK. A pretty bloody

simple problem even for that time, and Harris had opted for a very low cost demodulator from Data Control Systems (DCS), a vendor with several standard products for SGLS applications. It was a mistake in retrospect; the DCS unit had design flaws, but management did not have any indication at the time, and for that I am partly to blame.

To explain we must flashback to February, 1972. I was asked to run a bit error rate curve for a vendor demodulator on an overtime basis, since I had a full time assignment. A brief entry in my Engineering Notebook (#4563) page 99 dated 22 February, 1972 says "Trying to run test on DSC kludge." There is a set-up given, but no results. In my set-up the unit either ran error free or lost lock. I didn't document the threshold SNR. I told the Program Manager what I found, but apparently I was not convincing. He gave the job to a test engineer who subsequently claimed to have run a successful curve. I was busy at the time, and I was young and not all that confident, so I LET THE MATTER BE. I have offered this as a lesson many times over the years to junior people. If something doesn't seem right, feel right, smell right…it probably isn't right. Tell somebody! If that person doesn't resolve it, tell somebody else. Don't do what I did and let it go. Escalate any technical issue until it's fixed or you are shown to be wrong. If you're right, the problem is solved and perhaps you save the company a lot of grief and money. If you're wrong, you learn something.

I rubbed against this story again in the summer of 1974. It was a cafeteria lunch conversation with a couple of friends Ed Andre and Bill Roth who were Integration and Test engineers. They were leaving shortly for England to troubleshoot the Skynet Ground Station. The launch was scheduled for November, and there was a 2 dB shortfall in closing the link. Bill said, "I know where to find 1 dB." Bill and Ed thought the problem was "cockpit trouble," that the Brits were just doing something wrong. It didn't turn out that way.

I don't know exactly how, who, or when, but ultimately the decision was made to replace the demodulator, and they needed the replacement NOW. Jim Hays talked it over with a few of us in the modem group. The consensus was if management could get the "system" out of our way, we could do it. They agreed and a design team was assembled.

- Leader: Jim Hays
- Primary Architect: Ray Cobb
- Demodulator: Ray Cobb and Ralph Schaefer
- Bit Synchronizer: Wally Norris and Bob Smith
- Power Converter: Marty Levergood
- Mechanical: Bill Smith

It was a wild ride. The team settled into an ad hoc division of early people and late people. I was one of the latter, arriving late morning and working until 2:00 or 3:00 am. After that it was home for a few hours sleep, then back at it. Others preferred a more normal arrival time (7:00 or 8:00 am), and working until 9:00 or 10:00 pm. Everybody was working 12 to 14 hour days. We had our own lab and virtually carte blanche authority to pre-empt anything else that was going on. We borrowed or stole parts from whatever desk or cabinet had what we wanted. I think the only thing that went through procurement was a crystal oscillator. True to their word, management (and Jim Hays) kept the "ilities" folks mostly away.

The design was simple: A fixed 1 MHz LO converted the 1.024 MHz sub-carrier to a processing IF of 24 kHz. There was no AGC. Carrier recovery was by a squaring loop, with a voltage to frequency (V/F) converter as VCO. Data detection was by "filter and sample" using a single pole Low Pass filter followed by a comparator. Clock recovery was by a "slicer/filter" followed by a PLL whose configuration was similar to that of the carrier loop. Because of the low frequencies, all functions were implemented with operational amplifier circuits. Architect Ray Cobb says that an all digital architecture (being developed at the time by Jim Snell and Jon England) was rejected, considered too immature for the short schedule.

The packaging can only be described as crude. Circuit cards were made from blank copper-clad stock cut to needed dimensions. I don't think any two were the same size. Terminal posts were mounted on the card, one post per node, and were point-to-point connected with insulated wire. The boards were hard mounted via stand-offs to the inside of an aluminum chassis and interconnected point-to-point, no back-plane, no wiring harness. The single digital board was a "Standard Package" wire-wrap card, mounted into the chassis with stand-offs, the same as the analog cards. The front panel was painted solid gray, with a silk screened "sticky" affixed. Somebody (probably Bill Smith) found some standard front panel handles. A photo of the outside gives no hint of the tangled mess within.

Amazingly, it looked like we were going to pull it off. By Sunday the 10th, most everything was built, and it was all so simple that…well, not quite. In an understated "Apollo 13" style, Jim Hays wrote "we hit a snag, the bit synchronizer wasn't working right…"

The bit sync architecture that I originally planned was built around a commercial phase lock loop chip. It didn't work. These chips were designed for applications that required only frequency tracking. It was some time in the wee hours; Bob Smith and I were in the lab struggling with it. We tried several loop filter parameters, but it would not phase track. Finally I took the dikes and clipped the loop filter out of the circuit completely—no change! Bob and I just looked at each other with an unstated "what the hell do we do now?" I left. I don't remember whether I went to a bar or not, but think not, simply because there probably wasn't one open at that hour.

The next day I came in still unsure of what to do and was told the problem was fixed. Bob had re-designed it using a frequency scaled version of Ralph's carrier loop. I was appalled. He had wanted to do it that way earlier and I had said no. I didn't think the V/F converter oscillator could be made stable enough for the required bit sync loop bandwidth. Let's say we wanted a 1% fractional loop bandwidth for both applications (I don't recall the actual numbers). For the 24 kHz converted carrier that's 240 Hz, piece of cake. For the 250 bps data, 2.5 Hz., more like moose turd pie. I still think my skepticism was justified, but it seemed to work, and I didn't have another answer.

For better or worse the bit synchronizer was done by Monday the 11th. We did the unit testing, inspections, and wrap-up. In Jim Hays's files there is an undated bit error rate performance curve. The curve is corroborated by data in my Engineering Notebook #6125, page 68; the entry is dated 11 November, 1974. It ran about 2.8 dB from theory over a 1E-02 to 1E-05. I think the acceptance criterion was a single BER point at the expected minimum system receive level. On Wednesday the 13th the first unit was accepted by the customer. Turn-on was the 4th, so Wednesday was day 10, thus the moniker "Ten Day Wonder." On Friday the 15th the second unit accepted by the customer. By the end of the day both units and a power converter were packed and on a plane. Bill Roth, Ed Andre, and I took the units to England. The rest of the team took a well deserved weekend off.

I'd be amiss not to mention several other contributors. This list is from the article "10-Day Triumph," by Jim Hays in *Harris ink*.

Signal Processing Section: Jim Gann, Tony Campagna, Terry DeWick, Bob Trenner,
Bill Trent, Jerry Lorton, Jose Zayas, Paul Johnson
Support Organizations: Billy Cheatwood & group (Drafting)
Joe Dudley (Purchasing), Rea Hadley (Graphics)
Ron Carman (Photography)
Les Watson, Belle Confer (Quality Control)

I represented the design team at the site installation. I knew the design of the whole box, plus I was the only one on the team with a current passport. Once on site we put the system into a loopback test configuration and attenuated the input signal until the DCS demodulator lost lock. This established the system threshold with the DCS, and the attenuator was left at that setting. We then installed the Harris box. It ran error free at that level. Bill began to further attenuate the signal until the new demodulator threshold. Incredibly, the attenuator was turned down 18 dB!

To complete the story, two units were designed, fabricated, tested, delivered, and installed on station by 18 November, 1974. The launch of the Skynet Communications Satellite scheduled for November 19 actually occurred November 22. The three day slip was not related to this story.

By heroic effort, a mission-limiting problem was resolved. It was a barely averted disaster as I stated earlier; that's the bottom line. The RAF customer was mollified, but not happy. Harris had sold him a

system that wouldn't perform its intended mission. A barely averted disaster indeed, but it was averted. The following is my take on the success factors.
- Very modest requirements
- Design kept simple, partitioned for "divide & conquer"
- Competent & motivated team that knew each other well
- Utilization of parts on hand (including "stash" of individual engineers)
- Very long hours (including some "all nighters")
- Management support, rules bent where necessary
- "Drop whatever else you're doing" priority in support organizations

In summary:
"...it seems entirely reasonable to think that every now and then
the energy of a whole generation comes to a head in a long fine flash..."
-Hunter S. Thompson

The Vietnam Antenna
Jack Johnson
A 30 foot tracking antenna was built in 90 days to satisfy an emergency customer need in Viet Nam. Jim Perkins, Eddie Nelson and Ray Buchanan were principals on that project. I believe Mel Cox was the marketeer/program manager. Mel's not around to tell us anything, but maybe A. B. Amis has some recollections.

As I recall, it was a 30 foot dish, mounted on a Radiation "Telstar" pedestal. It is the only one I remember that was painted olive drab from top to bottom. (We structures guys like at least the dish painted white, to minimize thermal distortions). Eddie reported that the whole structure got so hot in the Vietnam summer, it was impossible to climb on it during the middle of the day.

Shipboard antenna replaced in 25 days.
From Radiation Ink
In November 1970, Radiation delivered to the Navy a Feasibility Model Shipboard Receiving Terminal consisting of two antenna pedestal assemblies mounted along the edge of the carrier flight deck.. One of the shipboard antennas was mistaken for an airplane drying rack by a pilot, and had to be replaced in 25 days. The antenna was severely damaged when an A-7 fighter overshot the flight deck of the 80,000-ton carrier U.S.S. Constellation during maneuvers at sea off San Diego and ended up hanging from the antenna.

Radiation was turned on November 28 to rebuild a replacement antenna to be ready for a sail date of January 5. The new antenna system was built in 25 days and shipped to the Constellation on December 21, due to the dedicated effort of five functional sections and several key individuals. This group includes Cliff Schoonmaker, Artie Ulmer, Russ Detherage, Art Sheldon, Lee Kalsch, Ron Pliego,
Homer Bartlett, Earl King, Dick Ott, Jim Crandall, Hank Farrow, Sid George, Bob Germaine and George Snead.

4.9 Radar Cross Section Measurements
Almost since the birth of radar in World War II, people have been concerned with how prominently aircraft and other targets appear on radar. Many factors affect this. The size and material of the aircraft has a big effect, but so do details of the shape and aspect at which it is viewed. The frequency and polarization of the radar is also important. An important step in designing an object to be "stealthy" is to understand the radar cross-section of existing aircraft as a function of aspect and frequency of the illuminating radar.

At one time, Radiation was a national leader in radar reflectivity measurements, probably because they had ample areas free of obstacles and interference. For some technology it helps to be

Aircraft hanging from Shipboard Terminal antenna.

located in the boondocks. Radiation's range was located at "80 acres", a then-remote area on the airport property, near the present location of the Harris headquarters building. Some full scale tests, using actual radars were also run from the Malabar facility. We were studying "stealth" before the term was invented!

The early company brochures prominently mention the facilities. From the 1960 Brochure:

Radar Reflectivity Facilities

Radiation Incorporated maintains a unique combination of facilities for the investigation of radar reflectivity characteristics covering a wide range of target types. The investigations range from the determination of the complex back-scattering matrix of simple geometric shapes to the evaluation of airborne active electronic countermeasures systems effectiveness. These facilities have measured values of effective back-scattering area from 100,000 square meters down to 0.000001 square meters. Objects measured vary from a single chaff dipole to an airborne B-52 bomber.

The facilities providing these measurements are of three types: (1) an indoor, low power CW facility for small objects, (2) an outdoor high power pulse facility for larger (but usually scale model) targets, and (3) a facility using instrumented modified military radars for dynamic inflight evaluations. All the systems produce relative measurements of echoing area To provide a calibration point for these measurements, a recording is made of a standard target before and after target measurements.

CW Measurement Facility

Radiation Incorporated has over 6000 square feet available in a modern laboratory for indoor measurements using small CW radars. These systems use separate transmit and receive antennas (or a hybrid junction) to achieve the transmitter-receiver isolation necessary to detect the echo return from targets. The echo signal is amplified and detected in a special receiver having linear dynamic range of 40 db. This detected signal is recorded on a pattern recorder in which chart paper motion is synchronized with target rotation. The target is rotated electrically in a mount having exceptionally low reflectivity.

Assorted antennas are available for measurement of radar reflectivity with almost any combination of transmitter and receiver polarization, (e.g., vertically polarized transmission with right circularly polarized reception). In addition the antennas can be widely separated to permit bistatic measurements where the radar transmitter and receiver are at different locations.

A combination of radar absorber material in large quantities and careful laboratory techniques facilitate the measurement of free-space values. An unusually good "microwave darkroom" is provided for these measurements.

The Pulsed Radar Scale Model Facility

A 300 acre outdoor range is maintained for radar reflectivity measurements of targets having maximum dimensions between approximately one and 15 feet. Four separate ranges are in operation, two at 24,000 megacycles, one at 10,000 megacycles and one at 3300 megacycles. One of the 24,000-megacycle systems is equipped for bistatic operation at angles up to 90 degrees in 10 degree increments and at ranges up to 2000 feet. The four systems are capable of operating simultaneously and provide automatic recording of echoing areas as a function of target rotation. For complex targets up to 30 feet of chart paper is used in plotting the echo area for 360 degrees of target rotation. The receivers and recorders have a 50 db dynamic range. The antennas in use permit measurements under diverse polarization conditions with control over both transmitting and receiving antenna polarization.

Narrow beam width antennas, short pulses, range gating circuitry, and low reflectivity model mounts are heavily relied upon to provide measurements of low reflectivity values free from the effects of background clutter. On the K-band systems, the return from a 50-cent piece is measurable 20 db above the background level at a range of 2000 feet.

These outdoor pulse facilities are used primarily for aircraft and missile reflectivity evaluation using scale model techniques, although full scale missiles and missile parts have been measured. Careful simulated scale factor choice permits measurements at frequencies ranging from about 75 megacycles through X-band for the medium size aircraft or missile configuration. The facilities have also been widely used in investigating the relative reflectivity of given parts of a target and in determining methods of reducing or increasing a given target's reflectivity.

The 1963 brochure:also has a section on:
RADAR REFLECTIVITY MEASUREMENTS

Radiation Incorporated has the most extensive radar reflectivity measuring facilities available in industry today, and has made more radar cross section measurements than any other group in the United States. The results of these measurements are contained in over 100 published reports Many of these are classified, but it can be said that the measurements have been conducted on a wide variety of aircraft, missiles, nose cones, nose cone decoys, tow targets, and enhancement reflectors. In addition, measurements have been made on items such as simple geometric shapes, absorber materials, artillery and mortar shells, and Echo balloons.

High power pulse measurements have been made at frequencies from 82.5 MC to 24 kmc with various polarizations and bistatic angles from 5 degrees through 180 degrees. While scale model techniques are most often used, full-scale measurements have also been conducted.

Low power CW measurements have been made at frequencies from L-band to X-band. Significant reduction of reflectivity has been made as a result of Radiation studies of aircraft, missiles, and nose cones. Our extensive capability and experience in this area also permits design of enhancement reflectors to increase the reflectivity of targets, decoys, and similar devices.

Other experience in this area includes correlation of static and dynamic measurements, correlation between theoretical and actual reflectivity, enhancement and reduction materials, and reflectivity data reduction and analysis. Significant advancements have been made in the art of static radar reflectivity measurements on large, heavy bodies having low cross section. Considerable effort is currently being extended in this direction.

Low power CW measurements have been made at frequencies from L-band to X-band. Significant reduction of reflectivity has been made as a result of Radiation studies of aircraft, missiles, and nose cones. Our extensive capability and experience in this area also permits design of enhancement reflectors to increase the reflectivity of targets, decoys, and similar devices.

Following are some of the targets measured by Radiation, Inc.

Aircraft		Missiles
8-58	Hustler	GAM-63 Rascal
B-57	Canberra	GAM-72 Green Quail
B-52	Stratofortress	GAM-77 Hound Dog
B-47	Stratojet	GAR-1 Falcon
F-100	Super Sabre	SM-62 Snark
F-94		SM-73 Bull Goose
C-47		TM-61 Matador
T-33		Missile A
Re-entry Vehicles		Miscellaneous
Nose Cones		Echo I Balloons
Nose Cone Decoys		Echo II Balloons
		Bistatic Reflectors
Tow Targets		Corner Reflectors Clusters Absorber Materials
ACP-1	TDU-6B	Simple Geometric Shapes
ACP-2	TDU-85	Artillery & Mortar Shells
MC-3	X-4	Light Aircraft Enhancement
MC-5	Q2A	Reflectors
PF4-FB8		Cylindrical Bodies

Numerous others could not be mentioned due to classification

4.10 AN/ASW-25

The ASW-25 was a data link for sending digital messages to aircraft. Its primary usage was to provide automatic landing capability for aircraft landing on aircraft carriers. It included a UHF Receiver operating on any of the standard communications channels. For automatic control, it provided both analog and digital signals to the autopilot. In addition, digital and analog signals were provided to drive displays for the pilot for pilot-assisted landings and other functions.

The initial development contract was received in and by 1971 over 2400 had been delivered.

Digital TV

In the late 1960's we received a contract from NASA-Houston for an experimental digital TV transmission system, using a combination of Data Compression and Error Correction. We called it DCEC or the Houston Test Set. This system used all the digital tricks at the time that almost all TV systems were analog. Our system demonstrated the capability to transmit a TV signal with 18 db less power than a conventional AM TV system. Now days your set top box uses the same ideas as our system of the 1960's which required three six foot racks of equipment.

Word Processing and Page Layout

Shortly after the merger in 1967, an effort was initiated to apply some of the Radiation expertise in digital processing to commercial applications, and one area selected was the pre-press area of newspaper publishing. At this time the state of the art in word processing was typing to paper with an IBM Selectric typewriter. In a few years, in the early 1970's, Harris 1100 "Editing and Proofing Terminals" were designed and installed at ToDay newspaper (Now Florida Today) in Cocoa, Florida. These allowed the entry and editing of text on CRT display terminals, and even provided for automatically type-setting the results, ready for printing in the newspaper. An article in a Cleveland newspaper of the time recounts to wonders of the system. A short time later, Harris delivered terminals which added the capability for page layout with graphic elements, for advertising and editorial applications. A considerable number of these systems were sold to newspaper and magazine publishers, but now Harris is little known in this field.

These were some of the earliest electronic word processing and desktop publishing systems. The first commercially successful word processing program, Wordstar, was introduced in 1976. Wikipedia states that desktop publishing began in 1983 with a program developed by James Bessen at a community newspaper in Philadelphia. In later years, Harris attempted other word processing systems, but they were never a commercial success.,

Early Programs

A B Amis offers some comments on some early programs. These are by JA numbers, which are chronological within categories (1xxx for CPFF jobs, 6xxx for fixed prices jobs.)

I believe 1007 was the UPM-19 test set, and I believe Ralph Johnson would have been involved with this project, which probably pre-dated my 1953 hire date.

1020 was an early attempt to make a solid state amplifier, by using using some sort of voltage sensitive capacitor as one of the tuning elements in an oscillator, and then hopefully recovering an "amplified" output from a discriminator circuit. A guy named Jim Jenkins ran this project back in old Bldg. 1 at the airport.

1021 was the original fixed stylus recorder, writing on maybe a six inch wide roll of Teledeltos voltage sensitive paper. It was intended to be used in flight testing aircraft, and I believe the Flight Test Lab. at WADC, headed by a shoot-from-the-hip guy named Jay Wayne, was the customer. Up to that point, analog strip chart pen recorders were the only choice, and they had obvious problems with G forces encountered in flight testing. Charlie Keith was the project manager, and it was being done back in old Bldg. 1.

1022 was the first project that Mel Cox and I were assigned to in early 1953, with Joe Detweiller as our technician. It was an attempt to develop an RF wattmeter superior to the Bird Termaline, which was the gold standard at that time in RF power measurement. Memory is pretty fuzzy on the "why" of this project, but my guess would be that the Termaline just metered the dc output from a crystal, which might not have too good a frequency response flatness, so what we were trying to do was to use a thermocouple to measure the temperature rise in a broad-band resistor vacuum deposited on a thin ceramic tube. We had a lab back near shipping, in the bottom rear of the then-new Bldg. 2 at the airport, and ran out of money before succeeding in coming up with the broadband ceramic tube resistor.

Artists concept of ASW 25 at work.

1023 was the development project on the AKT-6/UKR-1 PTM/AM telemetry system that was subsequently manufactured (only a few of them were manufactured) for use in Operation Teapot, where drone QF-80 aircraft (instrumented by the Instrumentation division in Orlando) were flown through the mushroom cloud from a nuclear blast in Nevada. Parker Painter, who had just come down from Melpar (where original design work on the AKT-6/UKR-1 had probably been done earlier), was the big cheese on the project, with Bob Dryden and George Anderson reporting to him. Some of the engineers working on the RF content under Dryden were Harry Siler, Al Campbell, and myself, assisted by technicians Don Graham and Ken Pohlman. Working under Anderson on the non-RF multiplex and de-multiplex content were engineers Dave Howard, Bob Bishop, and Bill Eddins, assisted by technicians that included Jim Danner for one. The project being done in a large lab area near the front of newly-opened Bldg. 2 at the airport in the 1953 and 1954 time frame. Helen Schmidt was our secretary.

1026. I'm pretty sure that this was an L-band radar receiver, with Ed Dorsett as PE and Jack Fifield as technician. Hans Scharla Nielsen may have been involved also in design of the resonant cavity tuner. This was being worked on in old Bldg. 1 in the 1953/1954 time frame.

1030 was the "fish" project -- a towed sonar device shaped like a torpedo -- that was ultimately lost in the Indian River, as I recall. I think John Spooner was involved with this project -- again in the 1953/1954 time period. George Shaw and Gantt Hamner both probably towed the fish behind their boats at one time or another.

1037 was a study contract for a wideband countermeasures intercept receiver, sort of looking at what was then the state of the art in broadband amplifiers like traveling wave tube amplifiers, chain amplifiers, etc. Vernon Dryden knew a little about TWT amplifiers from having worked with them in a lab at the University of Florida. Other folks on this project, classified Secret, were John Searcy, Jay Rosenvaig, and myself.

1038 was probably the original radar reflectivity contract from Bill Bahrett at WADC, and Mel Cox parlayed this 1953/1954 contract into ongoing reflectivity measurement work (at 80 acres) for many years to come. Some of the folks who later worked with Mel on reflectivity work were Wayne Williamson, Pete Young, Ronnie Evans, and maybe Dick Bivans. Modern day stealth aircraft would be an outgrowth of work done at the 80 acres facility.

1042 was probably the contract whereby Instrumentation Division in Orlando instrumented the QF-80 aircraft with various sensors and telemetry equipment. It was probably headed by Parker Painter, and I recall the Hans Scharla Nielsen was appointed something of a crew chief in some preliminary testing of the integrated system up at Eglin, and perhaps later also during the actual nuclear test. I know that Reed Barnett and Keith Taylor were among the crowd of people working on the project in Orlando.

1044 was a chaff measurement project, where a number of radars operating at different frequencies were set up as a test range down at the Valkaria airport off Minton Road and used over a period of years for measuring the radar returns from chaff (think Mylar icicles decorating your Christmas tree) dropped from aircraft over the boonies in South Brevard. This project would have been active in the mid/late 1950s, and I somehow associate the name Lee Miller with the project.

1047 was Radiation's entry into the PCM business that came to represent a major product line that continues to this day. Probably in the 1955 time frame George Shaw dispatched Don Graham and myself, armed with a camera and numbered adhesive strips, up to the Telemetry Lab at WADC to tear down an early prototype AKT-14/UKR-7 PCM/FM (probably developed by

Melpar) for shipment to Radiation. I'm sure someone else can remember the particulars better than I, but my recollection is that it was an 8-bit, 32 channel system, but I can't recall the clock rate. George Anderson was probably the overall PE on this job, as well as several other PCM contracts that quickly followed as an outgrowth of this job, and his projects occupied the whole of Bldg. 7 or 9, I'm confused about numbers, at the airport. After Anderson left to join the Dynatronics startup, Ralph Johnson inherited Anderson's PCM dynasty.

1055 was another contract for Jay Wayne at the Flight Test Lab at WADC, this one involving use of airborne PCM equipment identical to that being developed on 1047 to capture flight test data and record it on tape onboard the aircraft rather than telemetering it to the ground. The contract documents required that ground equipment also be supplied for playback of the recorded data for analysis, but Jay and Bill Dodgson had sort of had a "gentlemen's agreement" that the UKR-7 equipment slated for delivery to Jim Althouse's Telemetry lab at WADC would be used for this purpose, in the interests of minimizing our bid cost. This was all news to WADC's contracting officer, Lt. Hard, when the equipment was finally delivered (quite late, as were all PCM deliveries at this time), with no ground equipment!

1067 was sort of an interesting job from GE. The object was to see if circular polarization could help a K-band radar detect artillery shells during rain. I think everyone ended up with egg on their faces when it was realized that the shape of an artillery shell wasn't all that different from that of a raindrop, and the physics that made raindrops invisible would likewise tend to hide the artillery shell.

1082 As I recollect on this job we had a whole field of individual paint sample chips in a fenced-in area on the ocean at Patrick, exposing them to windborne salt spray, and our challenge was to try to duplicate those conditions using a new salt spray chamber in our environmental test lab. I somehow associate Phil Derrough with this project.

1097 turned out to be our "Golden Goose" in the mid/late 1950s, accumulating charges for various PCM jobs that had been intentionally underbid earlier in cornering the whole PCM field from competitors like Epsco. I believe the 1097 JA must have been sufficiently vague that nobody questioned the validity of charges against this job in making the other jobs "well" to the tune of possibly ten million dollars. I believe Charlie Keith was PE, and the ostensible purpose of the job was to develop telemetry suitable for testing ablative nose cones being developed by AVCO for use on ICBMs.

1104 was (I believe) the original TLM-18 60-foot telemetry tracking antenna -- which had been won by submitting an "alternate proposal" using large dish technology that we had been looking at in conjunction with a bid for a lunar radar for NRL. Vaudie Vice was PE, and I recall that John LaCapra was doing some mathematical analyses regarding surface tolerance on the reflector soon after he was hired in the late 1950s.

1105 was the Holloman Sled job, using PCM equipment to capture data from equipment being tested on Holloman AFB's high speed sled track. John Simmons was Contracting Officer at Holloman on this project before he came to work for Bill Dodgson in Contracts in the mid/late 1950s.

1130 was a Doppler radar navigation system for Sperry Gyroscope. I believe our hiring John Downs from Sperry was somehow instrumental in our receiving this contract award -- perhaps John wrote the spec at Sperry and then the proposal at Radiation? Warren Weiner was PE, and the job was originally staffed with a bunch of new grad engineers, that would have included Jay Fleming, Marvin Roth, and Milton Haff, and a few other more seasoned engineers including Guy

Pelchat, Bob Friedman, Jack Harrison, Dick Ferguson. The design was originally based on sub-miniature vacuum tubes, but a switch was made to transistors after Bill Eddins gave a few transistor design classes at the old Belcelona Hotel, which is now the site of Florida Air Academy. Walt Minnerly, Ed Dorsett, and I were temporarily assigned to the project in about 1959, and we all later moved out to the Valkaria airport facility before the project came to an inglorious end during flight testing (it really was working) when the landing gear on the test aircraft was prematurely retracted during takeoff, demolishing much of the system housed within a radome protruding beneath the aircraft.

1271 was a forward chaff launching system, using chaff-filled rockets launched from an aerodynamic pod hung underneath the wings of aircraft. Harvey Bush was PE.

1276 was for 28-foot antennas for Project Courier -- one of SatCom Agency's first ventures into satellite communications. These antennas used the same pedestals supplied by D.S. Kennedy, that the original TLM-18s used. Mel Cox sold the job and Dick Baker became PE. This job established a solid relationship between Radiation and the SatCom Agency that paid off in many subsequent contract awards.

Everybody knows about 1300's Hexagon Spy Satellite equipment now that the project has recently been declassified, but it was all top secret back in "the day".

1535 was for a six foot diameter K-band automatic tracking antenna system for receiving telemetry from the Dyna Soar [X-20] boost glide vehicle during the time when it was expected to be enveloped in an ion sheath as it re-entered the earth's atmosphere. The resulting antenna beamwidth was only 0.5 degrees, so acquisition of the re-entering craft was one major challenge. Boeing was the prime contractor for the Dyna Soar vehicle, and Radiation was a subcontractor to RCA, who was responsible for the K-band down link. Harold O'Kelley and I worked with RCA in submitting a successful "team bid" for the project, and the design of the antenna was complete and flight tested using aircraft prior to the Air Force cancelling the entire project. The project was active in the early 1960s, and space shuttle that came along several decades later was very similar in concept and appearance to the Dyna Soar -- just larger.

1769 was probably the contract where the "dielguide" antenna feed technique was developed and proven. Roland Moseley was PE, probably in the early 1960s, and the dielguide technique was used successfully on a number of subsequent antenna systems, including Mark V.

6098 was for an analog to digital to analog (sounds stupid, doesn't it) recording system for Douglas Aircraft at the Cape -- probably to provide quick look capability in static test firing of the Thor IRBM. The front end of the system would have been PCM, and the output displayed on a 640-stylus fixed stylus strip chart recorder. Ed Herrburger was PE, and the time frame was mid 1950s.

6151 was a contract from Sperry for a low noise receiver front end for a monopulse radar. As I recall, Lock Young was responsible for the design of a stripline mixer, and I was responsible for selecting matched mixer crystal pairs and designing a low noise IF strip with exceptional gain stability. Probably in the early 1960s. A number of these front ends were subsequently manufactured on other contracts.

Chapter 7

Radiation Printer

An outgrowth of Radicorder and fix styli recording was an ultra high speed computer printer for Lawrence Livermore Laboratory. The printer operated at 30,000 lines per minute, or about 7 pages per second. The next fastest printer of that era I have found was the Stromberg-Carlson SC-5000, which was rated at 5000 lines per minute. The Radiation printer output was not the highest quality, but it was fast. In the right situation, the technology and/or the experience could have been leveraged into more computer business, but Radiation had other business to pursue with their limited resources.

There is an interesting account of this printer from the users perspective below.

The Radiation Printer

by George Michael

Editor's note: This is an expansion of an article that appeared in CORE 1.4, a Computer Museum History Center publication, in November, 2000.

There are very few computer users left who still can recall the frustration of having to wait for a printout. For instance, around 1953-1954, at the Lawrence Livermore National Laboratory (LLNL), the first printers used in conjunction with the UNIVAC I--our first computer--were nothing more than typewriters with print rates of perhaps 6 characters per second. Since the typical output from a design calculation involved 50,000 to 100,000 characters, printing would take an inordinately long time. The quest for speedy printing at LLNL led us through a succession of interesting machines, one of which we relied on for about 10 years, starting in 1964. This was the so-called "Radiation Printer", an eccentric and demanding invention that met our computer printing needs for speed despite its own oddities.

To better appreciate what the Radiation Printer represented, it will be useful to consider first, the few printers that preceded it. I've already mentioned the use of suitably modified typewriters on the UNIVAC I. The IBM computers that arrived next had modified 407 accounting machines serving as printers. They were rated at 150 lines per minute, a line being up to 132 characters wide. They were also far too slow to meet our needs. As I recall things now, IBM never included faster printers for its initial computers. If you dealt with IBM, these 407s were what you got; a Hobson's choice.

ON THE JOB HEARING LOSS One of our first attempts to get something faster than these printers arrived from Remington Rand around 1957. This was a 600-line-per-minute impact printer, where a line included any number of characters from 0 to 120; each page held about 50 lines. As fast as this was, it was still too slow to serve the needs of dozens of people who spent too much of their valuable time waiting for results.

Also, when these so-called impact printers ran, the noise level was dangerously high A few intense users lost some of their hearing from standing in front of the printer, anxiously trying to read their output as it was being printed. In addition to being very noisy, impact printers were not sufficiently reliable, so we sought other solutions.

THE GIRL WITH A CURL We tried a marriage of cathode ray tubes and xerography: The SC5000 built by Stromberg Carlson in 1959. This device formed characters by projecting an electron bean through a character mask, creating a spatial distribution of electrons then formed the selected character when plotted on the screen of the CRT. The SC5000 further selected where to position the character along the print line.

The light thus generated was projected onto a selenium-coated drum that is fundamental in the xerographic printing procedure. In this process, after the image was formed on the selenium drum, it was dusted with xerographic powder ("toner"), which adhered only where the light had suitably charged the surface.

By bringing paper in contact with the drum, the image was transferred. The paper then moved through an oven where the powder was fused to the paper. Input to the printing system was via magnetic tape.

The SC5000s were modified so that they printed at an impressive rate of about one page per second. This required expanding the fusing oven and adding a Rube Goldberg device to z-fold the printed output. Quite often, the paper would catch fire as it moved through the fusing oven. The printer kept running, but now acting more like an automatic stoking device, feeding fresh paper into the fire! The SC5000 was very much like the angelic little girl with a curl right in the middle of her forehead: "when she was good, she was very, very good, but when she was bad, she was horrid."

THE RADIATION PRINTER Even when printouts were produced at the one-page-per-second rate, the total time was just too long to meet the aggregate needs of all users. The search for faster printing continued, so everyone was primed to welcome a new printing technology, ultimately embodied in the so-called "Radiation Printer".

Two technologies came together in the Radiation Printer. First, the actual print process was based on an electrographic printing technology, and second, the process was wedded to a standard printing press that far predated the advent of computers, but was rugged and reliable. Before the arrival of computers, most printing presses were designed to produce many copies of the same page. For LLNL applications on computers, the problem was to produce just one copy for each of thousands of output files. The electrographic technique, which is both fast and clean, uses light to carry information to an electrically charged material where a toner was used to make the image visible. The image is then transferred to paper where it is fixed by chemistry or heat. Xerography is a good example of this technology. Even though further discussion of the process is beyond the scope of this article, some basic differences as used in the Radiation Printer are important to note.

Instead of light, electronic charge was used to carry the information. The charge was made to produce an electric arc from a selected stylus to a black electrographic web through a whitened paint-like material that coated the web.

The arc burned a tiny hole in the coating thereby revealing the blackness of the web. This made toning and fixing steps unnecessary. One saw a black dot, and enough black dots produced a simulacrum of the image sent by the computer.

This type of printing process was normally used for the production of mailing labels for magazines like Time and Newsweek. Although no actual printer existed, everyone felt confident that a printer could be scaled up from a mailing label size to a larger page format, and it seemed it could be made to go quite fast and it promised to be economical. We solicited bids for a high-speed printer, and what became known as the Radiation Printer was chosen.

Some salesman got the Radiation Printer to brag about itself. Here is a quote of some of what it said (I omitted some parts that were inaccurate): "The Radiation Incorporated (more of this later) Printer operates in a line-at-a-time mode, providing 30,000 alpha-numeric lines per minute, each line containing 120 characters. The input data rate of 60,000 characters per second is compatible to (sic) the data transfer rates between many existing digital computers and magnetic tape output units. Automatic transfer between the magnetic tape units allows for nearly uninterrupted data flow into the printer.

"Key to the printer's high speed is its Electrosensitive, Multistylus Recording Technique which eliminates the mechanical inertia of high-impact mechanisms and permits a dry, immediately available output without subsequent processing. High-speed recording is attained by swiftly moving the recording paper under a closely-spaced row of fixed styli. Styli are selected according to the character to be printed and energized with high velocity current pulses. Passing these pulses of current through the electrosensitiverecording paper exposes high contrast marks on the paper. A paper transporting system handles the paper so that the printer need not be interrupted to add paper..."

The printer had 600 styli arranged in 100-styli modules. The print area was about eleven inches in width, and the page was eleven inches tall. The images were not considered to be very high resolution. A traditional printing press was used to move the web past the styli. The procedure was dubbed "Revelation Printing," because the coating was burned away by the charge coming from the styli, thus revealing the black paper underneath.

An annoying problem with the styli arose during operation; they tended to get contaminated with burnt paint debris, and therefore stopped functioning. The solution had nothing to do with modern technology. Cleanliness was achieved by blowing pulverized walnut shells against the styli. It was claimed that other nuts would not work.

A few additional remarks seem to be in order. First, the Radiation Printer had nothing to do with radiation, but simply was named for the company that built the printer: Radiation, Inc., of Melbourne, Florida. The company modified a real (Hamilton) printing press and added the needed electronics and controls to produce a printer that ran at seven pages per second (for Indy drivers, this turns out to be about 4.3 mph.) Printers in the newspaper business run even faster although they don't seem as versatile. In addition to printing at that speed, it punched binding holes at the top and bottom of each page, perforated each page so jobs could be separated, fan-folded the output, and separated the jobs one from another. The various performance numbers for the printer are summarized in the tables below.

There were enough styli to allow up to 120 character positions per print line, and each character was formed within a 7 x 9 dot matrix. Suitable spacing between characters and between lines of characters was thereby provided, so that in practice, a page could contain up to 10 columns of numbers each up to 12 decimal digits, each column containing 55 to 60 numbers. The capacity of a page was thus about 5,000 characters. It was also possible (but not easy) to address any point in a line, so that with some special programming tricks, graphs could be produced. Printing was thus accomplished exactly as a video-scanned raster is produced.

Something in the print process gave the output a disagreeable odor. Some of the users actually complained of headaches. An investigation of the odors failed to expose any serious health hazards, so the simplest response to this was to authorize the use of fans that could keep the odors away from those sensitive noses.

THE IMPLICATIONS OF SPEED So what does seven pages per second mean to the users? Each page was approximately 11 inches square. This implies the speed of the paper through the printer was about 77 inches per second. The print data was supplied from any magnetic tape able to provide a nominal 60,000 characters per second--we used IBM 729 tape handlers written at 800 characters per inch. Such tapes had a nominal rate of transfer of up to 62,500 characters per second, more than adequate for printing, so the extra time available allowed for the filling and emptying of buffers, and for the movement of the paper past areas at the top and bottom where no printing was done. On balance then, of the 7 pages per second, about 1.3 pages-worth of that time was not used for printing, but for the extra movement of paper required to get from one page to the next, as well as time for hole punching and page scoring. Thus the rated speed of 7 pages per second meant that the user was getting about 6 completed pages per second within that 7-page time. This printer minimized user waiting time or it serviced more users in a given interval. Very few of the users (physicists) could read at this speed. In effect, then, the throughput speed of this printer was generally adequate to meet the needs of the growing user community, and it did so for a bit over ten years.

The Radiation Printer was integrated into the normal operations of the computation department, and very quickly was producing around forty million pages per year. This was only about one-fifth to one-third of its capacity, which was a good thing. The machine could be taken down for emergency maintenance, and still very quickly clean out the entire print backlog when it was brought back on line. Later on during its tenure, some microfiche recorders were added. Their annual output quickly grew to about 130 million pages distributed over about one million pieces of fiche. The effect on the Radiation Printer was less than expected however: The annual output dropped to around thirty million pages per year and stayed there. For most users, the fiche was used for long-term storage of their computational results, and output from the Radiation Printer was used mostly for day-to-day checking. When a project was finished, the paper was generally discarded.

CONCLUSION The output from the Radiation Printer was hard to read; the gray-on-black paper was heavier than ordinary paper; it had, for some, an undesirable odor; and it took up too much storage space. The output was not pretty; the users often referred to the printouts as "scunge," but it met their needs, producing at the rate of 7 pages per second. None of the printers that were brought in to replace it ever came close to this speed. However, as effective as the printer was, no one shed a tear when it was removed sometime during the late 1970s.

AFTERWARD It's always humbling and sometimes instructive to ask if anything was learned.

There are several lessons available, though who learned them is not clear, nor is the question of whether the lessons have had any long-term positive effects.

Somewhat in the spirit of a post mortem here are some things that were learnable:

Simple works best soonest.

Speed wins – most of the time.

True zealots will put up with practically anything to get the job done.

On the matter of print quality, most users prefer quality more than they prefer quantity.

In the course of dealing with users of all sorts, we evolved an additional rule to help get through the day: Generally, if somebody doesn't know what to do, don't ask them.

TABLE 1. Approximate pages of computer printouts per month in 1978
   Teletypes 200,000
   35 mm film 600,000
   On line Printers 830,000
   6 Microfiche Recorders 9,800,000
   Radiation Printer 3,400,000

TABLE 2. Early computer printing to 1974 (Approximate speeds)
Typewriters 1/20 Lines/sec 1953
Line Printers (IBM 407) 2.5 Lines/sec 1954
High Speed Printer (Rem. Rand) 10 Lines/sec 1958
SC500060 Lines/sec 1959
Radiation Printer 420 Lines/sec 1964

TABLE 3. A Summary of Radiation Printer Performance Numbers.
Print Technology Electrographic, Revelation
Data source Magnetic Tape, up to 800 bpi; 75 ips.
Character Rate up to 62.5 Kcps
Print Rate 7 pages/sec; 4.3 mph
Print size 5000ch/page

A NOTE ABOUT DATES: More precise dates may exist, but most official records are in a state of flux. The dates used here are my best approximations. Similarly, the values in the Tables are extracted partly from several unpublished internal reports. They are intended mostly for comparisons.

For information about this article contact: George Michael – gmichael@attbi.com

One of the cabinets shown is the SC5000, circa 1960, [predecessor to the Radiation Printer] that was prone to catching the paper on fire. The print rate was one page per second, with input via magnetic tape. One can also see a homemade device for Z-folding the paper.

Operator Mona Millings stands at the table where the separated output was delivered from the printer, and at the other end, the large rolls of paper used by the machine. Paper from the rolls could be spliced head to tail so there was no need for rethreading through the press. A roll lasted about 45 minutes and a special dolly was needed to move the rolls, since at over 200 pounds, they were far too heavy to be moved by hand.

This photo shows the back side of the printer.

Rolls of paper waiting to be fed into the Radiation Printer

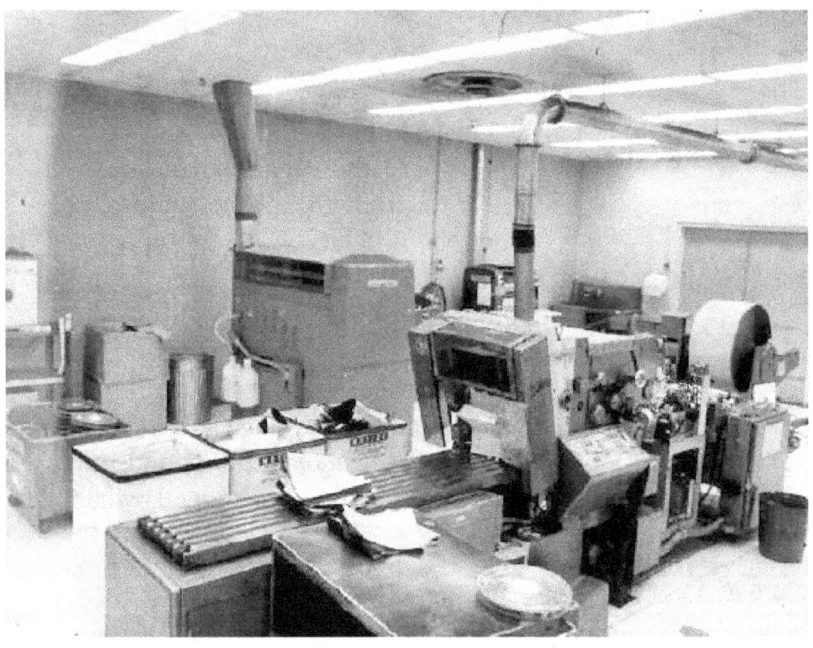

The machine page-perforated, hole-punched, fan-folded, separated the print jobs, and deposited them onto a slowly moving set of belts.

The paper required for the Radiation printer was a sandwich of a black conductive layer coated with white top layer. The overall appearance was a blueish gray. Printing was accomplished by an electric arc burning a tiny hole in the white coating to reveal the black layer underneath. Too bad we can't reproduce the odor here!

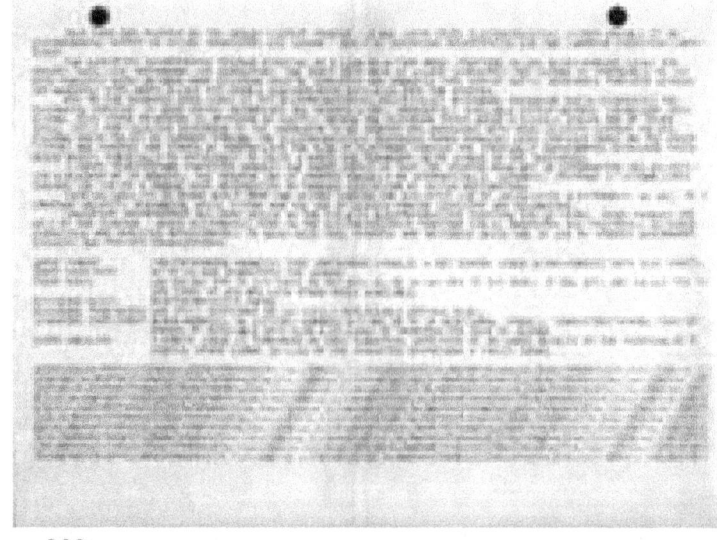

Chapter 8
Stonehouse Tales

Stonehouse and Bayhouse were two programs managed by Radiation, Incorporated, from 1963 through 1975. Many people worked together on these programs, sometimes intensely. Close bonds were created that have lasted for nearly fifty years. These tales have been assembled to chronicle some of the adventures shared by these people. Hopefully, they will prove amusing to others as well.

Asmaritis
By Jack Johnson

I'm sorry if this story sounds a little indelicate. I'll do the best I can, and hope that you will accept it with the same good humor as those of us who lived it.

The Italian army occupied Ethiopia from 1936 to 1944, and brought in many civilians during that time. Many of those civilians settled down, and were allowed to remain after the army was driven out. This group formed a merchant class that still existed at the time of the Stonehouse/Bayhouse project. During those twenty-five years, they had introduced into Asmara a lot of amenities we take for granted.

Among those was a water treatment and delivery system. Most houses and businesses had running water. Unfortunately, there was a flaw in the system that had not yet been corrected. Sewage was usually routed out of town in open ditches, and the water supply mains were frequently buried in these same ditches. Consequently, the water was severely contaminated with every pathogen imaginable. Some of those pathogens were potentially lethal, so this was a serious issue for expatriates. For those who had experienced both, "Asmaritis" was ten times worse than a bad case of Montezuma's revenge.

Kagnew Station tried to mitigate this problem by treating its water with massive additions of Chlorine. Like many rural towns, the Kagnew skyline was dominated by a large water tower that stored and distributed this water. It was drinkable and relatively safe, but it tasted absolutely awful. Vegetables cooked in this water came out snow white, and had practically no taste at all.

For those expatriates who lived off-post, which included all of us, access to this treated water was essential. Every family had a five-gallon plastic can, and we were allowed to fill them from a faucet under the Kagnew water tower. Every day, someone from the family living quarters, and from the Shrader hotel, would load up all the cans and make a water run to Kagnew.

Despite the many precautions we took to avoid contact with untreated water, Asmaritis proved to be inevitable. We had to use that water for bathing, washing and flushing. For dish washing, we added our own chlorine, and then used Kagnew water for rinsing. We learned that there remained a lot of ways to accidentally ingest a drop or two of untreated water. That was all it took. No one could reasonably expect to get through their tour without at least one substantial case of Asmaritis. As far as I know, nobody did.

Aside from the illness and dehydration,, Asmaritis presented obvious logistical problems. Our family members stayed close to their toilets, but the workers on site weren't so fortunate. At the beginning of the job, there were no amenities on site at all. We had to dig a hole and build our own two-hole outhouse. It naturally had to be downwind, but we had to place it much nearer than we would have liked. We tried to locate it within the range of a reasonable walk. For people working at ground level, that was usually (but not always) sufficient. If one was working up on a structure, however, it was another matter. By the time one could climb down, walking was frequently not an option. Many didn't make it. To cope with the frequent accidents, we had to develop our own version of savoir-faire.

I'd better not go into further detail, or else this story might become x-rated or juvenile. I hope it hasn't already. If you want graphic details, make up your own. Chances are, if you can imagine it, it happened to at least one of us.

Black Friday
By Jack Johnson

This story isn't particularly funny, but it does show that sometimes it's better to be lucky than good.

The 85-foot Stonehouse antenna had two separate drive systems for moving its big dish around. Its primary and operational system was a servo-controlled hydraulic drive system, with hydraulic pressure provided by large, 440-volt, three phase pumps. Its secondary, or "aux" drive, was manually driven, with 220-volt AC motors, driving through an additional drive gearbox. The hydraulic drives were necessary for tracking and precision pointing of the dish, but were much too complicated and labor intensive for routine maintenance applications. Both had their uses, but obviously could not be used at the same time. The hydraulic motors could remain mechanically coupled all the time, but the aux drives had to be decoupled before operating with the primary system.

During the late fall of 1964, the Radiation team was operating two shifts on site. The night crew was primarily involved with mechanical tasks, and used the "aux" drive system to slew the antenna around to various positions. The day shifts were working on the checkout phase on the various electronic systems, including the hydraulic drives. Facilitating the shift change required decoupling the aux drives and making the necessary wiring changes to permit closing the servo loop around the hydraulic drives. To streamline switching back and forth, Chris Catsimanes had devised a temporary patch plug, made from a 68-pin connector, which could be substituted for one of the cables at a junction box on the antenna structure. With the plug in place, the aux drives could be operated. With the plug removed and the cable in place, the hydraulic system was operational.

This switchover procedure worked well until a Friday morning, when someone forgot to remove the patch plug. The morning shift fired up the big 440-volt generators, turned on the hydraulic pumps, and all hell broke loose. We didn't know it at the time, but the jumper plug closed a sneak circuit that routed 440 volts throughout the entire system!

Within seconds, the site was abuzz with frenzied activity. It was reported that Gerry Edwards made footprints on the walls of the equipment room, racing around throwing circuit breakers. Elsewhere in the same room, Eddie Bannister was standing in front of an electronics cabinet. Eddie was pulling out circuit boards one at a time, examining them, and pitching almost all of them over his shoulder into a growing pile of burned-out boards. Standing behind Eddie was Tony Barile, smoking his curve-stemmed pipe. Except for dodging the flying circuit boards, Tony appeared to be completely calm. Inside, he was frantically trying to figure out how to recover from this disaster.

Next door in the control room, Joe Pepin was sweeping up glass from the Nixie tubes that had exploded out of the console. Chris Catsimanes was under the console, examining what was left of the servo control electronics.

This was serious business! The customer had ordered lots of spare parts, but practically none of them had yet arrived on site. Most were in various stages of manufacture back in Palm Bay. Recovery could take months. Aside from the potential impact on national security, Radiation had a fixed-price contract and would have to absorb a lot of cost.

One small thing was on our side. The time difference between and Asmara and Palm Bay gave us all day Friday to assess the damage before the home team got to work. At least we could save one day if we got some help from home.

By the end of the Friday shift, we had prepared a list of the parts we needed, and sent it on its way to Palm Bay by Telex. Even so, we thought it was a long shot that the message would get through in an intelligible fashion. The telephone lines ran all the way across Africa before connecting to the "real" infrastructure. Messages, if they got through at all, were usually garbled to the point they had to be repeated. Anyway, it was a start.

All we could do on Saturday was to make whatever repairs we could, and look for damage not already discovered. There was no joy in Mudville that night! Were it not for Chivas Regal, the weekend would have been even more dismal than it was.

On Monday morning, Walt Chick arrived at Asmara, as had been scheduled. The only thing unusual was the inordinate amount of luggage he had brought with him. Walt had been at home, preparing for the trip, and had not been briefed in any detail on the disaster we had experienced. All Walt could say was, "Kent Busing told me to bring these extra suitcases to you guys". Jim Crenshaw had to call for a second Volkswagen to carry Walt and the luggage to the site.

When we opened the extra suitcases, we found every one of the items we had asked for only 65 hours before! By the end of the day Monday, the Stonehouse system was completely operational again.

None of us could understand how the Palm Bay team had grasped the situation so quickly, and had been able and willing to work around the clock to get all this stuff together and ready to go. But there it was. The class 6 store at Kagnew Station still had some Chivas Regal, so we were all able to drink a toast to the Palm Bay heroes who had pulled our chestnuts out of the fire.

Needless to say, the offending patch plug was never used again. Had it not been fried in the incident anyway, it would surely have been cast into the darkness of the Eritrean wilderness.

Civil Engineering Isn't Always Boring
By Jack Johnson

To design and build a foundation for a heavy structure or building, one needs to know a great deal about the earth underneath. You can't just dig a big hole and look around, because the foundation needs to sit on undisturbed soil. What is needed is a soil profile and analysis, based on boring samples taken at various places around the foundation location.

Fred Heartz, a professional civil engineer and soil analyst, designed the foundations for virtually all of the antenna foundations for Radiation. Fred had designed and built his own rig for drilling and taking core samples, and had it mounted on the frame of a relatively old truck.

His Radiation contracts always included both the soil analysis and the foundation design. This required shipping his boring rig to the construction site and back for each job. That old truck had lots of miles on it, but not many on the odometer.

The Stonehouse customer didn't want us shipping things to the site ourselves, so we had to let them handle the shipping of Fred's rig by Government Bill of Lading. Although it must have been very expensive, they had it airlifted to Asmara by the Military Airlift Transport Service (now called "MAC"). It arrived safely, and Fred soon had his core samples. He turned the rig back over to MATS, and returned to Palm Bay to design the Stonehouse and Bayhouse foundations.

Fred's boring rig should have arrived at Patrick AFB soon after Fred. When the expected arrival date came and went, Fred began to get worried. He needed the rig for other clients, some of them Radiation programs. MATS put tracers on it, but couldn't tell him anything definite. He asked Radiation for help, and George Lane volunteered to represent him. The customer told him to file a claim, and he and George did.

Fred didn't really want money compensation – he wanted his rig. Self-employed, he was out of business until he got it. The customer didn't want to give up on finding it, either, so neither side was in a hurry to give up and settle the claim.

After about three months with no progress, they agreed that it was time to settle up and move on. Fred got enough money to build another rig, plus some compensation for his lost business revenue until he got it into service. The matter seemed settled.

About two years later, the customer called and said the wayward rig had been found, and was on its way home. Apparently the MATS loadmaster had, for reasons never explained, decided to offload the rig in the middle of the airfield at Djibouti. It had been sitting there ever since. The waybill information was deliberately circumspect, for security reasons, so this strange-looking package was not easy for the local MATS people to identify. It just sat there until it got in somebody's way enough to put it back into the system.

The Stonehouse customer didn't try to get their money back, but shipped it to Fred anyway.

Fred wasn't sure he needed a second rig, but he knew from experience that one might get lost in the mail.

Don't Listen to Your Mother about Underwear
By Jack Johnson

Only two years prior to the Stonehouse/Bayhouse installation, The Ethiopian Government had canceled a Federation agreement with the country of Eritrea, and had seized and annexed that country. Eritrea was now a part of Ethiopia. This had not been an easy or friendly assimilation. The peoples of both countries had different cultures, different traditions, and very different languages. We didn't fully realize it at that time, but the stage had been set for a growing insurgency and eventual revolution. The dissenters were regarded as bandits. They were called "Shiftas"

The Shiftas, for the most part, hid out in the countryside, outside the population centers. Their primary objective was to collect weapons, to arm themselves for the coming conflict. They did this largely with the tactics of the stagecoach robbers of the Old West. They would stop vehicles on the road, look for weapons, and take anything else of value, including the vehicle. The treatment of the victims depended on their nationality. Ethiopian nationals rarely survived the encounter. Neither did the Italians, especially if weapons were found. They had nothing against the Americans, and, except for robbing them, treated them well and let them go. Even so, it was better not to have weapons in the vehicle.

When the authorities occasionally captured Shiftas, they were brought into town and hung in the public square. Their bodies were left hanging for three days, as a deterrent to potential sympathizers.

There wasn't a lot to do in Asmara, and the Americans at Kagnew Station weren't content to stay on the post all the time. The Army built R&R centers at Keren and Massawa to give them somewhere to go. Providing armed escorts for every traveler wasn't a practical defense against the Shifta problem, but they kept careful track of their people on the road between centers. All travelers were registered at a military checkpoint when leaving a location, and looked for at a calculated time at the destination checkpoint. If you didn't arrive on time, they came looking for you.

The Radiation employees at the site all worked at least 60 hours per week, and seldom got a chance to go to Keren or Massawa. But their wives did. Any time they could commandeer one of the Volkswagens, they would leave the kids with a volunteer and hit the road. They did observe one precaution, though, based on the following incident:

Among the many expatriates working in Eritrea were a number of Peace Corps volunteers. All were young and fearless, as were we. One day, three of them, two boys and a girl, took a trip out in the wilderness of the Eritrean plateau. Their intent was to visit one of the many native villages in the region. They were accosted by Shiftas and held, while their car was being searched. Not finding anything of military value, the Shiftas merely took everything they had. Then they left them standing there, naked as the day they were born. Three days later, they were found, wandering around the desert, by a group of natives from a nearby village. They were none the worse for wear, except for being hungry and having severe sunburn in places infrequently exposed. The natives help them get home safely, and their story circulated far and wide.

Knowing this, the intrepid Radiation wives always cautioned their embarking travelers, "make sure you wear your oldest and scruffiest underwear, and maybe the Shiftas won't want it!"

"Going Down!"
By Jack Johnson

Among the many things shipped to Asmara were two brand new P&H 100-Ton cranes. Since the Bayhouse 150-foot dish was welded together on the ground, these big cranes were needed to lift it into its place on its pedestal.

Either of these cranes, when outfitted with all of its boom and jib segments, could reach above the top of a 20-story skyscraper. When not involved in the "big pick", these cranes were used individually for conventional lifts, and for transporting ironworkers and tools to the top of the structures. Most experienced ironworkers, and some Radiators, would choose to "ride the headache ball" just above the crane's hook. For larger groups, and those with tools and parts, an 8-foot square platform was used as a makeshift elevator. This device was a wooden skid with waist-high sides, rigged out with a four-cable bridle attached at each corner.

This elevator was a perennial favorite with visitors to the job site. Raised to its highest level, a person could experience a panoramic view of all of Asmara and much of the surrounding plateau. Some even claimed to have seen the Red Sea to the East.

Besides that, the trip up and down was a thrill in itself.

Many VIP's back at Palm Bay had expressed a desire to visit the site when the construction had advanced to the point there was something to see. By the late summer, 1964, both antennas were in place, and the system integration had begun. There was something quite impressive there to see.

At this point, a visiting entourage was arranged. Included in the group were Homer Denius, our Board Chairman, Dr. Joe Boyd, our recently hired President, and Carmen Palermo, our new Chief Technical Officer. George Lane, our Contracts Administrator, escorted the group.

Upon arriving on site, our visitors were given the usual routine, consisting of shiny new hard hats with their names on them, and a walking tour of the site. In addition to hard hats, they all had cameras on lanyards around their necks. Inevitably, one of them spotted the "elevator", and asked if they could go up and take pictures. That was easily arranged, and they all climbed in. The crane took them up to the tip of the jib.

In the high steel community, there is an initiation rite for "greenhorns" riding the crane for the first time. Unfortunately, we all forgot to tell the crane operator that, although these were certainly greenhorns, the initiation would best be saved for another time.

When our visitors finished viewing and taking pictures, they signaled they were ready to return to earth. At that point, the crane operator completely released the cable brake, and the elevator went into free fall.

As the elevator descended at a terrifying pace, both the occupants, and those whose paychecks they signed, saw their lives flash before their eyes. It was frightening just to behold. The occupants seemed to be floating above the elevator, their cameras floating above their heads, and above both, hard hats.

About 30 feet above the ground, the crane operator jammed on the cable brake, and another spectacle took place. First, the approximately 500 feet of cable began to stretch like a rubber band. That, plus the flexure of the crane boom, allowed the elevator to continue downward under heavy deceleration. That sent occupants, cameras and hard hats crashing into the bottom of the box.

About 10 feet above ground, the descent stopped. But the show wasn't over. None of us had ever heard of bungee jumping at that time, but we got a demonstration anyway. The box and its contents bounded back upwards several times before settling down at about 20 feet. The operator then lowered the box gently to the ground.

Nothing happened for a while. Then we saw fingers rising over the side rails. Slowly, ash gray faces began to appear. Not a single hard hat was to be seen. Eventually, everyone was accounted for, and our new initiates began to climb over the rails to terra firma.

We all thought our Radiation careers were over, but we had overlooked one important thing: VIP's, so long as they are male, also have the "never admit you're lost or scared" gene. Not a single person complained, and several agreed that, "All elevators should work that way". All claimed they wanted to do it again, but maybe at some other time. In the meantime, they were more interested in the direction to the outhouse.

That night, any remaining apprehensions were assuaged at a party we had for them. Dr. Boyd demonstrated his skills at line dancing, and Homer Denius had a sporting go at the Limbo bar.

"Maybe Tomorrow We Pour Concrete"
By Jack Johnson

One of the most critical and tension-producing milestones on the Stonehouse/Bayhouse schedule was the completion of the concrete foundations for the two antennas. Unless these huge underground structures were complete on the day that Rohr's ironworkers arrived, the project schedule would be irretrievably impacted, and much unnecessary cost would be incurred. Pres Shrader had contracted with a local Italian construction firm to do the work.

The sequential steps in building an antenna foundation are as follows: excavation; rock drilling; surveying; formwork; rebar installation; anchor bolt installation; concrete pouring and vibrating; curing; form stripping; backfilling and compacting; and more curing. Different people and equipment had to be on site for each of these steps, so a lot of planning and scheduling were required to get it done correctly and on time.

The Italians were neither lazy nor stupid – they knew how to do the work. But their sense of urgency was no match for us Americans. They didn't care much about our schedule, or theirs, for that matter. To them, "domani" was just as good as "oggi" for getting things done. Their answer to any question about progress was, "Maybe tomorrow we pour concrete". They said this, even though it was obvious that they were weeks away from being ready.

By the time I arrived on site, they were clearly falling behind. Pres was beside himself with worry. Escalating our concerns through their management didn't help a bit. You guessed it – they said, "Maybe tomorrow we pour concrete". Pres and I had to micromanage the whole thing. We didn't get off to a good start, as each of us made surveying mistakes that slowed things down even more. Eventually, though, we managed to get each foundation to its "pour day" in time to support our deadline.

Pouring concrete into a monolithic foundation is a symphony of activity, and must be choreographed to perfection. Each foundation required hundreds of cubic yards of concrete, and the wet concrete had to be poured in one continuous operation. It all had to be done in a single day, with no stopping to fix something forgotten. This was made even more difficult by the primitive and limited equipment available. There we no batch plants, concrete delivery trucks, no chutes. All we could get were about six small mixers, the largest of which was about one third of a cubic yard. With these, dozens of wheelbarrows, and hundreds of Eritrean unskilled day laborers, we would have to do everything on site. Pour day was going to be a long, hard day.

Nevertheless, the day came when we had the formwork ready, the mixers laid out around the perimeter, the piles of sand and gravel, the bags of cement, and the tankers full of water. Then we could all say with some confidence, "Maybe tomorrow we pour concrete!" (weather permitting).

The next day, we let the Italians manage the work, and every one of the Radiation team (and even the customer reps) became quality inspectors. We had somebody watching every move, and every one had the authority to call a halt if required. The concrete was being continually mixed in batches in the six mixers, with laborers loading sand, aggregate, cement and water into one end and other laborers hauling concrete out of the other. Wheelbarrows moved everything.

The Eritreans would dump their barrows into the forms where told. As this was happening all around the foundation forms, other Eritreans were down inside the forms vibrating the concrete to eliminate voids. This was continued until the forms were full.

One critical process requirement was to take concrete samples from the mixers from time to time. These samples were tested for correct water content, and when cured, for strength. These tests absolutely had to be passed, as there was no way to repair the foundation afterward. It was also necessary that these samples be representative of the whole pour. We had to provide assurance that the samples were not taken from a batch that had been "doctored" in any way. Our strategy for doing this was to have the Radiation inspector assigned to each mixer decide when to pull samples, and to point at the mixer and yell "that one!" Upon hearing that yell, all Radiation personnel who could be spared would rush to the mixer and completely surround it. That was to make sure that the mix got no further ingredients.

We never flunked a single sample test, and many strength tests were exceeded by a wide margin. Somehow, the Italians would figure out how to throw an extra bag of cement into the batch, right under our eyes!

We heard the phrase "Maybe tomorrow we pour concrete" so often that the procrastinators among us adopted it and took it home with us. For several years after that, program managers at Palm Bay, upon inquiring about a schedule, would be puzzled by the rejoinder, "Maybe tomorrow we pour concrete!".

Mr. King Harold
By Jack Johnson

We always needed lots of manual labor around the site. Pres Shrader had negotiated with the elders of the nearest village for all of the Eritrean helpers we would need. Any one who had a job to do could assemble himself a crew of these laborers and go to work. The Eritreans didn't know a word of English, and our guys couldn't even make the babbling sounds of Tigrenian. Communications were a matter of arm waving and demonstration. One could not have predicted this, but two of our team proved to be natural leaders in this style of leadership. Arlie Wheeler was one and Earl King was the other. This story is about Earl King. The 85-foot Stonehouse antenna structure sat on six large steel legs. The foundation supporting these legs looked from above like a giant concrete block, a squared figure eight with two open cells in the middle. Four of the legs were attached to the corners of this block, and the other two in the middle of the sides.
Once the foundation concrete was cured, and the forms stripped from it, we had to backfill and compact the space in the two cells, and pour slabs over the top. We needed the weight of the compacted soil to help stabilize the antenna. Besides, we didn't want two big holes under the antenna.
We had several large air compressors on site, and lots compacting tools. These tools were similar to jackhammers, except that they had a rounded steel foot on the end of the shaft. Most tools consisted of a single air cylinder, but some had as many as three. The three-legged machines had handlebars for control, and their operation was quite comical to watch. Held completely level, all cylinders would operate, each with its own rhythm. If the apparatus was tilted, the downhill cylinder would stop, and the uphill cylinder would double its hammering pace. The laborers learned how to operate these devices, but only a few could master the "three-legged devil."
The backfilling procedure consisted of dumping about six inches of earth into the bottom of one of the foundation cells, wetting it a bit, then compacting that soil to a firm and smooth surface. While compacting one cell, other laborers would, using shovels and wheelbarrows, add the six inches of soil to the other. Then the two crews would swap cells and repeat the cycle. This activity would continue until the ten-foot deep cells were filled to within 4 inches of the top.
The average Eritrean laborer weighed about 100 to 110 pounds, soaking wet. Hauling the dirt and operating the compactors was about all they could manage. But they were willing and enthusiastic. They also had the leader they had chosen and respected. He led by demonstration, and was the most powerful man they had ever seen. They thought he must surely be the overlord of all of the foreigners on the site. They knew all about Kings, but they had never heard of an Earl. To them, he was "Mr. King Harold". Whatever Mr. King Harold required was not questioned.
As the backfilling began, it was discovered that the crew swapping took quite a bit of time. Ladders had to be lowered into the cell so the workers cold climb out, then the ladders moved to the adjacent cell to let them climb down. The equipment was then hoisted out of one and into the other. Earl was not patient with this arrangement for long, so he started reaching down and fetching his workers out by hand, then lowering them into the other cell without putting them down. He soon discovered he could handle them two at a time. Not to let such an awesome display go unappreciated, the worker team redoubled their own efforts, and pretty soon the dirt was flying in both cells all the time. There was nothing the team wouldn't do for their "Mr. King Harold".
Still, I'll bet they all slept well that night. Including Mr. King Harold.

Security, Radiation Style
By Jack Johnson

We at Radiation thought we knew all about classified programs. We had done a number of them over the 13 years of our existence. However, when Stonehouse, immediately followed by Bayhouse, came along, we found out there were security levels we never knew about. For one thing, we didn't have a vault. We didn't have time to build one, either. The only thing we could do quickly was to secure a portion of a building we already had. The west end of Building 1B was sealed off for the project, and we had to change badges to go through the only door. (This door separated the project from the restrooms, so we all had to put nature's call on hold while we exchanged our inside badge for our outside one.) The rest of the RF department, along with all other contracts, was housed in the East end of the building. Roland Moseley referred to himself and all of those people as "S.L.O.B's" (Sorry Left-Over Bastards).

Shortly after we had the area secure and the people inside, it was time for our official program security briefing. Bryant Helmer, our security chief, came over and climbed up on a table in the lab area. Everyone crowded around while he told us what was expected of us. Everything was on a "need to know basis" and was not to be discussed openly, even within the "vault". At the highest level of protection were the mission and capabilities of the project. Any unauthorized disclosure of these items would likely result in the termination of the program. Close behind those was the location of the installation. We were admonished to report any violation,, or even loose talk, about these areas. We weren't thrilled about the prospect of squealing on our close friends, but we accepted that this was really important. We were even told we could not ship anything to the site, because the documentation could be used to reveal the site location. Even the names of the program, "Stonehouse" and "Bayhouse" were sensitive, and the job should be referred to simply by its internal JA number.

A number of us, including the Helmers, belonged to a duplicate bridge club. We met about once a month, and rotated from house to house. It was Bryant's turn to host the next event, shortly after the security briefing. As Anna and I were getting out of the car in front of the Helmer house in Magnolia Manor, his wife yelled from their front porch, "Hey, I hear y'all are going to Asmara!"

This situation left me with two questions I was never able to satisfactorily answer: (1) "To whom do you rat out the Chief of Security?, and (2) "To whom do you address question (1) ? " After agonizing over this for a while, I chickened out and merely reported it to Bryant. I never heard any more about it.

A few months later, the people slated to go to the site were getting ready. We had to get lots of inoculations, and Radiation arranged for us to get them from Dr. Omainski. As I walked into his office, he said, "Hey, I hear you are going to Abyssinia!" (Abyssinia is the biblical name for Ethiopia.) It seemed like everyone in Melbourne knew about the location of the project.

A few years after the program was completed, we were told by one of the customers that there had been a full day in 1964 that the program had actually been canceled.

Drew Pearson was probably the best known (and most feared) columnist in America at that time. One day, while the government ship carrying the Stonehouse/Bayhouse materials was at sea, Pearson published an article in his "Washington Merry-go-Round" that disclosed almost every sensitive thing about the program. There had obviously been a huge and deliberate security leak. The customer agency was appalled, and made an immediate decision to cancel the program. They ordered the supply ship to reverse course and return to port. During the ensuing 24 hours, they continued to assess the damage of this disclosure, and finally decided that the program was just too valuable to scrap, even if our adversaries now knew about it. The ship was ordered to turn around again, and to continue its course to the port of Massawa. The customer didn't tell us about any of this, so we were blissfully ignorant of the close call we had had.

Shortly afterward, though, we did notice that there were a lot of sneaky characters living with us in Asmara's Ciao hotel.

The Great Putty Ball Fight
By Jack Johnson

The Stonehouse antenna, like most steerable dishes, was driven by large, semi-circular gears on each axis. These gears were much too massive to be shipped and installed in one piece. They were cut up into smaller segments, and carefully adjusted back into place during construction. When their alignment was completed, the space remaining underneath the segments had to be grouted to prevent further movement. The grout used was a very strong liquid epoxy, which had to be held in place while curing. To accomplish this, Rohr temporarily dammed up the edges of the segments with a thick, sticky, mastic. This mastic was thicker, but otherwise similar in consistency to Play-Doh. Once the grouting was complete, the mastic served no further purpose. But it didn't hurt anything, either, so Rohr left it in place when they departed the site.

It was a calm and pleasant afternoon. Everyone was busy, as usual, but there were none of the usual crises to deal with. One Radiation engineer, working out of our "cherry picker bucket", was doing something near one of the big drive wheels. He suddenly had an idea. Without thinking about where it would inevitably lead, he pulled off a small handful of this mastic, rolled it into a ball, and bounced it off the hard hat of another nearby engineer. Not one to let this affront go unchallenged, the victim immediately responded and fired back.

Like any good bar fight, this exchange quickly erupted into a free-for all. Many working on the structure climbed to highest point having access to an ammunition supply, and started heaving putty balls at anyone within range. Even people on the ground joined in, picking up putty balls and returning them into the battle. The instigators and principal culprits in this shenanigan included Jack Davis, Pres Shrader, Chris Catsimanes, Jim Turner, and of course, me.

Not everyone participated, but they all thought this was hilarious and very entertaining. Good hits were cheered, especially if the victim was one of our "big shots". Even the native laborers stopped and enjoyed the fun. They thought Americans were crazy anyhow, and this display proved it.

We all knew this was going to lead to trouble, but that didn't seem to matter at the moment. Dick Wilson, our site manager and resident father figure, got out the bullhorn and called the above-named culprits into his office. We were going to get scolded, and everybody knew it. The non-participants thought this was hilarious and very entertaining, too.

Dick lined us all up and reminded us that our behavior was not only akin to running with scissors, but was unseemly for us current site leaders and presumptive future corporate executives.

Thus ended the great putty ball fight. But until the last of the mastic was gone, everyone kept a watchful eye on the usual suspects.

But I don't care what Dick Wilson said. That was fun, and I'd do it again. Right now. All I need is a putty ball, and cherry picker, and someone with a nearby hard hat.

The Hard Way Construction Company
By Jack Johnson

In addition to the two large antennas, Stonehouse and Bayhouse, the Asmara installation included two guyed boresite towers. Each of these towers was triangular in cross section, three feet on a side. The near field tower was 200 feet tall, and located on the job site about 800 feet east of the antennas. The far field tower was 400 feet tall, and located on the top of a small hill about two miles west of the site. We had contracted for these towers with E-Z Way Towers, Inc., Clearwater, Florida. It was a turn-key contract for the towers, and we would mount our boresite antennas and run the cabling. The tower materials had been delivered to the site, concurrently with everything else.

Everything was in readiness for E-Z Way's assembly of the towers. We had completed the necessary site work, and had installed the concrete foundations and "deadman" anchorages. At that point, for reasons never fully explained to us, the government decided not to grant the security clearances required for E-Z Way's crew to travel to the site. We had to immediately go into a workaround mode (a position not unfamiliar to us). The 200 foot antenna would not be a problem. Its 40-foot sections could be assembled on the ground, it guy cables attached, and lifted into place in one piece by one of our 100-ton cranes. The 400 foot tower was another matter. Us young, brave immortals wanted nothing to do with it, and we knew Radiation Management wouldn't let us do it anyway. Dub Hudson would have had a fit if he knew we were even thinking about it.

Let me digress for a minute, and explain why this task would be so daunting. First, there was no way to get our crane to the site. Anyway, the tower was not strong enough to be lifted after horizontal assembly. It would have to be erected, one 40-foot section at a time, without a crane. Second, the guy wires were attached in 100-foot intervals, meaning that the partially assembled tower would have up to 100 feet of unstabilized height at any point. The first 100 feet had no guys at all, and would have to be temporarily lashed in place until the first three sections had been installed.

The lifting of the sections would have to be accomplished with a winch and gin pole. This was the really dangerous part. The gin pole itself was a small triangular tower, about two feet on a side, and about 80 feet long. At its top was a frame supporting a three-foot rotating boom with a pulley on its end. This device was called the "rooster", and it resembled a hoisting davit used on boat docks. A steel cable ran though this pulley, and down to a winch on the ground at the base of the tower. The assembly procedure called for strapping the gin pole to the previously installed tower section, such that about 50 feet of the pole protruded above the top of that section. The next tower section was then hoisted by the winch to the top of the gin pole. One of the reasons this step was very dangerous was the winch itself. An AC motor powered it, and its speed was too fast and uncontrollable. Movement could be accomplished only in jerks. There was therefore a high risk that the lifted section might snag against the tower or guys, and that could bring down the tower and the two or three people that were required to be on the tower. Only keeping the rising section away from the tower with long tag lines could mitigate this risk.

When each tower section arrived at the top of the gin pole, the most frightening step of all came next. For the lifting, the "rooster" obviously had to extend away from the tower centerline. Before the new section could be lowered into place, a brave soul would have to climb to the very top of the gin pole, and literally kick the "rooster" around 180 degrees to center the pulley above the tower. The operation sometimes occurred as much as 150 feet above the nearest guy wires. At this height, the flimsy tower and even flimsier gin pole were literally flapping in the breeze. And the tower section hanging on the cable could go pretty much where it pleased.

After each new section was secured in place, another dangerous step had to be accomplished, called "jumping the gin pole". For this step, a pulley was attached to the top of the tower itself, and the winch cable routed through it and attached to a point near the bottom of the gin pole. The cables fastening the gin pole to the tower were loosened (not removed), and the gin pole lifted 40 feet to its new position. The danger in this step was that two or three climbers had to be inside the tower, constantly unjamming and relocating the fastening cables. All workers would have to remain in perfect coordination with the winch operator, or..(well, I won't describe all the bad things that could happen).

Now, back to the narrative. The Rohr-contracted ironworkers heard Pres Shrader and me discussing this dilemma. After a huddle, some of them said, " If you can wait until our work for Rohr is finished, and if you will pay us a lot of money, we'll put those towers up for you!" This sounded very attractive, but we were worried that Radiation would still be carrying the risk. George Lane, our contract administrator, happened to be one the site at this time, and he offered a novel idea. He said to the ironworkers, "If you wish, we can charter a new corporation around you, and Radiation can give you a contract to do the work. I have the know-how and the authority, and we can do it here and now. All you have to do is decide who's in it, how you will share the risk and the proceeds, and who will be the corporate officers." They agreed to proceed, and Delaware got a new corporation. Naming the company Hard Way Construction was a natural. I'll bet the union chief of the Roanoke local of the Ironworkers Union would have had a conniption of he ever learned that one of his members was a CEO on the side.

Thankfully, no one got a scratch doing the work. Still, we couldn't bear to watch.

The Shrader Hotel
By Jack Johnson

One of the most difficult logistical problems of the Stonehouse/Bayhouse project was providing housing for the fifty-odd individuals and families scheduled to be there by the summer of 1964. Kagnew Station was there, but it was a tiny military base, whose guesthouse had only four rooms for overnight visitors. There were no available houses or apartments in Asmara, because there was no growth generating demand for them. Pres Shrader was sent to the site, months in advance, to work on this problem. To provide for the families with children, he eventually had to hire an Italian contractor to build an apartment complex. For the ironworkers and unaccompanied Radiation employees, he had to find something else.

While driving around downtown Asmara, he noticed a large, apparently abandoned, three-story building. It didn't appear to be for sale or lease, so it took a day or two to find out who owned it and what was in it. When he finally got inside, he found what appeared to be an old hotel. It had many rooms, and lots and lots of plumbing (and a few other things I will mention later). The property needed a lot of renovation and cleanup, but seemed otherwise ideal for our needs. The owner was willing to provide the renovations and sign a lease, all for a reasonable price. A deal was quickly struck, and we were in business. The building was known thereafter as the Shrader Hotel (and maybe still is).

A month or so later, as the early members of the Radiation team were getting ready to depart for the site, one of them was approached by Larry Kjerulff. Larry had served in the North Africa campaign during WW II, and he had a request to make. In his typically quiet voice, he asked, "During the war, there was a legendary brothel in Asmara, known all over the Horn of African. It was called 'The House of Mirrors'. Would you see if you could find out what happened to it?" He wouldn't explain how he knew about it, or why he cared. Nevertheless, we said we would check it out for him.

I suspect you've figured out where this story is going. Sure enough, once we knew the questions to ask, it didn't take long to confirm that the Shrader Hotel and the House of Mirrors were indeed one and the same. Strong clues were in the building itself. In addition to an unusual amount of plumbing, many of the walls (and ceilings) were covered with mirrors. Most of the remaining walls were painted with lurid murals. All of these things were carefully preserved in the restoration. Apparently when the Italian and German soldiers had all been driven out, and the allied soldiers had themselves left, the brothel operator merely closed the door and walked away.

Months later, we were told, by some of the Italian old timers, that we had done a great job of restoring the old brothel (and not just the building) to her prior glory. The ironworkers and crane operators, not widely known for their chastity, loved the place and didn't want to leave. My wife would never let me go down there, so I never found out why. The unaccompanied Radiation people who stayed there also admitted they liked it and learned a thing or two. But like Larry Kjeruff, they wouldn't say why.

"What Did You Expect, Tivoli Gardens?"
By Jack Johnson

People who were traveling to Asmara typically departed on an Ethiopian Airlines Boeing 707 from Athens at midnight. After a fueling stopover in Cairo, the plane would arrive in Asmara about 8:00 AM. The Cairo stop was always unpleasant, as Gamal Abdel Nasser and his United Arab Republic was hostile to all Europeans and Americans. People who got off the plane were frequently held at gunpoint until the plane was ready to re-board. Those who stayed were subjected to airport officials running up and down the aisles and spraying everyone and everything with bug bombs.

The passengers on the May 27, 1964 flight were the usual complement, except for one thing. In addition to the Americans, Europeans and occasional Ethiopians, the cabin contained about 15 ironworkers from the Roanoke, Virginia local of the Ironworkers' Union. They were the first wave of workers hired by Rohr to build the Stonehouse and Bayhouse antennas at Asmara.

The ironworkers raised the usual amount of hell on the flight, and everyone on board was pretty annoyed with them by the time the plane landed in Cairo. However, after soundly booing the airport officials and their bug bombs, they all settled down for the remainder of the trip.

Most of the passengers were visiting Asmara for the first time, and some had never been anywhere in Africa. For their benefit, the pilot made an extra effort to show them the countryside. Just before the final approach to the Asmara airport, he banked the 707 in a slow arc for a nearly 180-degree turn. Everyone looked out the window at the craggy mountain peaks and valleys below. The rainy season had not begun, and there was not a trace of vegetation or animal life anywhere. The engines were idling, and no one said a word as they studied the utter desolation below. The cabin was very quiet. In a minute or two, the soulful voice of one of the ironworkers was heard to say, "Mercy, mercy, what have I done?"

The drivers and greeters who met the plane that morning saw something they had never seen before. Every one of the passengers and crew were still laughing when they left the plane. A most unusual response.

Why Surveyors Aren't Afraid of the Dark
By Jack Johnson

This isn't a funny story, but I hope it is an interesting one. I threw it in for the amusement of those who remember why they wanted to become engineers.

The Stonehouse antenna had 21-bit Grey Code shaft encoders on each of the axes of its pedestal. We needed that resolution to provide the precision RF beam pointing required for its mission. I mention this to set a standard for the level of accuracy we had to achieve in all of the alignments we would have to make in setting up the system. All of the optical measurements would have to be made with accuracy in the low, single-digit arc second range. We didn't want to waste a single one of those 21 bits. Hundreds of such measurements were made – far too many to describe here. This story is about a representative one of them.

The arc-second standard challenged the capabilities of our K&E theodolites. Surveying anything in the daytime was out of the question, because of solar thermal distortions to the structure. We had to do our surveying after dark, and after the structures cooled down to ambient temperatures. (The latter didn't take as long as you might expect, in the thin air of the 8000-foot high Eritrean plateau.)

One of things we had to determine with such precision was the bearing of the line of sight between the 85-foot Stonehouse dish and the 8-foot diameter calibration dish on the boresight tower. The 400-foot tower was located on a hilltop about two miles away, and the calibration dish was mounted right at its top.

The problem was that the test dish was unlit, and you couldn't see it in the dark. The fact that the tower was almost due west gave us an idea. As the morning sun rose, the first thing it would illuminate was that dish! If we could survey it at just that moment, we could get a very accurate bearing before everything started to heat up.

Al Fritz and I went to the site about an hour before sunrise to get set up. We had already embedded a brass monument in the concrete precisely under the antenna, and had established a very accurate north-south baseline through the mark on the monument. We leveled up the theodolite and its tripod over the mark, measured the height from the mark up to the scope's pivot, and waited. Suddenly, there it was! The distant calibration dish was lit up like a full moon, yet everything else was still pitch black. Perfect!

The procedure for maximum accuracy was to take six sets of data, throw out the one or two with the biggest deviations, and average the rest. Between sets, we needed to verify that the instrument was still level and centered above the mark. We were going to have to move fast to get all of this done before the rising sun started moving things around.

Things went well for five sets, and then suddenly everything went crazy. The instrument wouldn't stay level, and the Azimuth and Elevation readings of the target were much different than before. While I was focused on looking through the instrument, Al said, "take a look at the tripod!" Although everything else was still dark, there was a one-foot segment of one leg lit up with sunlight. Something in the structure above us had become sunlit, and had reflected its light down onto that leg. That little bit of indirect sunlight had been enough to begin expanding that leg, throwing the instrument out of alignment.

Fortunately for us, the five sets of data we already had were tightly packed, so we could confidently average them to get the coordinates we sought. All we had left to do was correct the coordinates for the vertical offset from the theodolite scope up to the center of the big dish, and convert the new coordinates from Az-El to X-Y. But that was just mathematics – the engineering was done.

Chapter 9

Branches, Acquisitions and Spin Offs

It is interesting to examine some of the major organizational changes, voluntary and involuntary, over the history of Radiation.

Very early in the game, several divisions and organizations were established in Orlando, an hours drive from headquarters in Melbourne. This is an inconvenient distance for close co-ordination, so why did this come about?

One of the early Orlando organizations was the Research Division, started in 1957, presumably because the founders felt the need for PhD's to lend credibility to the new company. It was probably easier to recruit such people to Orlando than the small town of Melbourne. However, to reap internal benefits of such an organization, close, convenient contact is necessary. Radiation benefited much more when, years later, the Advanced Technology Division (ATD) was established in Melbourne with the charter to use half their resource in direct support of the development divisions.

Another early Orlando operation was the Instrumentation Division, founded in 1954 to implement an important program to install instrumentation in jet aircraft. The Melbourne airport's runway would not accommodate the jet aircraft and efforts to get Melbourne to extend it failed.

The program (Operation Tea Pot) succeeded, but the Instrumentation Division needed follow-on work to grow and prosper. Unfortunately, the interests and capabilities of Orlando Instrumentation Division were very similar to those in Melbourne. Such competing divisions are always a problem. since engineers always think their way is better than others. This resulted in, among other things, similar but not identical logic cards, defeating any advantages of scale. It is also difficult to mange the marketing and bidding by the divisions so that they are not destructively competitive.

The final Orlando Division was the Production Division, opened in 1953. This made more sense, not being directly competitive with any Melbourne operation, and drawing from a larger labor pool than was available in Melbourne.

The disadvantages out weighed the advantages, and all Orlando operations were finally closed in 1963.

Early on, Radiation felt there was a need for more presence in California, since many of our potential customers were located there. In 1959, Radiation acquired Levinthal Electronic Products, reportedly when Homer met Elliott Levinthal on an airplane flight, and he seems just the type of person to hit it off with Homer. The story leads to a long devious path, but is interesting to follow. Come along.

Levinthal was born in Brooklyn in 1922. He graduated from Columbia in 1942, received a master's degree from MIT in 1943 and a PhD from Stanford in 1949. During World War II, Levinthal worked with Sperry Gyroscope, where he earned a patent for the discovery that detuning the intermediate cavities of multi-cavity klystron amplifiers increased their efficiency, an important development in klystron design.

While at Sperry, he developed relationships with faculty members in Stanford's Physics Department, and these friendships prompted him to continue his studies at Stanford. As a doctoral candidate he worked under the direction of Nobel laureate Felix Bloch. Levinthal's dissertation, on the magnetic resonance of the hydrogen atom, was part of Bloch's Nobel-winning discoveries in magnetic resonance.

After completing his doctorate, Levinthal joined Varian Associates as one of the founding employees, rising to serve as research director and, ultimately, as a director of the company. Varian purchased the rights to the Bloch patents and Levinthal was responsible for developing them into commercial instruments, laying the groundwork for use of nuclear resonance as a tool in chemistry and biochemistry.

n 1953, Levinthal founded his own company, Levinthal Electronics Products, which developed some of the first defibrillators, pacemakers and cardiac monitors. The company also built high power UHF transmitters, including the 430 MHz unit for the Arecibo Observatory in Porto Rico capable of 2.5 MW pulses, and 150KW in the CW mode, and the FRC-39 Troposcatter Communications System. The company also built other high power electronic devices including energy storage banks and tube testers.

Levinthal Electronic Products was combined with the Space Communications Division, renamed Radiation At Stanford, and sold in 1963 to Energy Systems, Inc. (of which I can find no record).

The Radiation Space Communications Division was also founded in the near by Mountain View area in 1959 and built primarily digital systems, most notability a Satellite Control Facility for Vandenburg AFB, and the ill-fated Data Acquisition Cart (incorporating the Radiplex Low Level Multiplexer) that they called Radatac. This was eventually finished in Melbourne. As mentioned, they were eventually combined with Levinthal to form Radiation At Stanford.

Back to Dr. Levinthal, who had a varied career after leaving his company. In 1961, he joined the Genetics Department of the Stanford School of Medicine, where he worked with Joshua Lederberg on exobiology, examining the question of extra-terrestrial life and designing experimental missions to Mars. This led to Levinthal's service on a number of NASA and National Academy of Sciences committees. He became a member of the photo interpretation team of the Mariner 9 Mars Orbiter missions and deputy team leader of the Viking 1976 Lander Camera Team.

At the medical school, he became a research professor and director of the Instrumentation Research Laboratory. The laboratory was active in the design and operation of mass spectrometers and their computer control and analysis. As associate dean for research affairs at the medical school, he developed computer systems for the Medical Center, which led to the creation of SUMEX-AIM, the Stanford University Medical Experimental Computer for Artificial Intelligence in Medicine.

During a three-year leave from Stanford, he served as director of the Defense Sciences Office at DARPA, the Advanced Research Projects Agency in the Defense Department. His responsibilities there included materials science, geophysical sciences, electronic science and system sciences. He also introduced programs in the biological sciences and robotics.

Returning to Stanford, Levinthal became a research professor in the Department of Mechanical Engineering and director of the Stanford Institute for Manufacturing and Automation (SIMA). He later became associate dean of research at the School of Engineering.

In addition to his academic roles, Levinthal was active in the development of a number of Silicon Valley companies. He was involved in Arthur Rock's investment partnership, the first venture capital fund in Silicon Valley, and served as director, science consultant and science adviser for several companies dealing with medical or biological questions. He was a founder of Neuroscience, later renamed Eunoe, which pursued research on devices to filter cerebrospinal fluid to treat Alzheimer's disease.

Levinthal was an active and generous philanthropist. He provided financial support to less advantaged students to attend Stanford and was a generous supporter of the humanities. His philanthropy included the establishment of the Levinthal Tutorials program in creative writing at Stanford University. Levinthal was also active in local and national politics. He had a great love of travel, leading him to all seven continents in his adventures. He died in 2012 at the age of 89.

Micro Electronics Division History

One of the interesting and different of tangents taken by Radiation was a venture into semiconductor design and manufacturing.

First, some background. Rudimentary semiconductor diodes were used in the 1940's. The first transistor came into being in 1947, but the first large scale use was in transistor radios introduced about 1954. The first computer using transistors was the CDC 1604, first delivered in 1960. The first integrated circuit, using a handful of components, was invented in 1958. Small Scale Integrated circuits (SSI), using 10's of transistors were first delivered in about 1962 . Medium Scale Integration (MSI) , using a few hundred transistors came into use in the late 1960's. LSI (Large Scale Integration) , using a few thousand transistors or more, was introduced in the mid 1970's.

At Radiation, some of the earliest systems, even for airborne use, used vacuum tubes. By 1959, when I went to work there, we were using individual transistors in all our designs. and soon SSI came into common use.

It soon became clear that integrated circuits would have a huge impact on our systems, particularly air and space borne systems. At first they were viewed primarily as the next generation of compact packaging, a key technology to our business success. Radiation management felt that we needed an in-house capability in this technology to maintain our competitive advantage in systems with stringent space, weight and power constraints, the heart of our business.

It turned out that the importance to a government systems contractor (like Radiation) of an in-house source of IC's was vastly over-estimated, and perhaps was even negative. IC's were far more than an advanced packaging technique; they were a new technology, expensive to design and implement, but with huge impact in improving performance and reducing production costs (per transistor). The IC branch of a systems company needs heavy internal usage of their products by the Systems Branch to justify the development costs. Upper management seeks to achieve this by mandating to Systems the use of the internal IC's. Systems designers resent externally dictated design choices and have powerful arguments against them, using data they can generate. ("They make my system perform worse and cost more.") External sales by the IC Branch could achieve the necessary sales volume to support their large expenses, but sales to competitive systems companies (particularly of a break-through product) negate the reason they were established to start with. Large scale sales of commodity devices to the commercial marketplace is so different from the government systems market that there was little if any value to having the IC operation part of a systems company.

The history of Radiation's adventures in the IC business is not well recorded , and was outside the experience of most old Radiators, who were involved in the systems business. The best record is in articles in Radiation Ink, the newsletter, from which we draw heavily.

The first mention of an integrated circuit organization, the Physical Electronics Division (PED), was in the August, 1964 issue. Uryon (Urie) Davidsohn is listed as head of this division.

A July 1965 article announced the first product, a programmable matrix of 50 diodes. Programming was achieved with fusable links; this appears to be a very early use of this technique. The product was announced by Trygve Ivesdal, manager of Product Marketing and was displayed at the March 1965 IEEE show.

A June, 1966 article showed the organization as follow:
Jack Hartley, VP
Product Engineering, Don Sorchych, Director
Process Engineering, Dr. Don Mason, Director
Operations Support Department, Dick Grey, head.
Manufacturing, Tom Callahan, Director.

In October, 1966. PED was renamed Micro Electronics Division (MED)
Also in 1966 Don Sorchych was appointed Director, Product Development
The November 1966 issue announced that MED had moved into Building L (now Building 4)
June 1967. MED backlog announced as $1.5M.
January, 1968. Don Sorchych appointed VP-General Manager of MED, replacing Hartley
February 1968. Article on RA-909 Op Amp appeared, listing Trygve Ivedal as MED Product Marketing Manager. Copy of RA-909 ad shown.
April 1968. Lamar Clark named Engineering Director of MED, succeeding the promoted Don Sorchych.
September 1968. VP/GM Don Sorchych announces appointments at MED:
    Skip Bliss, manager of Material
    William Boecklen, Manager, Cost Accounting
    Peter Brooks, Manager of Manufacturing Engineering's
    Donald Cook, Sectio Head, Device Design
    Donald Fritsma, Manager General Accounting
    Harry Nystrom, Midwest District Marketing Manager

> John Sarace, Section Head, Materials Development
> Monroe Weiant, Section Head, Digital Circuit Development
> William White, Section Head, MOS Circuit Development

January 1969. William Weir, MED Marketing Director, announced that Charles Merrick was promoted to National Sales manager for MED

March 1969. It was announced that MED is to display an off-the-shelf line of Radiation Hardened circuits at IEEE show. The products include a family of 930 series DTL, a 709 Op Amp, a dual level shifter and a dual 4 input line driver.

June 1969. MED building under construction, north of Palm Bay Road.

February, 1970 William Weir, Marketing Director, announced that the price of RA-2600 Operational Amplifier sliced by 50%.

July 1970 Raytheon presents award to MED for Poseidon IC's.

March 1971 Hamilton Standard presents award to Harris Semiconductor for IC quality.(First use of Harris Semiconductor name.)

July/August 1971. Sorchych announces appointments:
> Len Ornik, V_P Operations
> Jack Kabell, Director, Research and Development.

May/June, 1972. Sorchych announces promotion of Richard Gray to Director Programs and Product Assurance.

July /August, 1973. HarrisScope, Semiconductor Division newsletter, begun.

Wikipedia states: "In 1988, Harris acquired GE's semiconductor business, which at this time, also incorporated the Intersil and RCA semiconductor businesses. These were combined with Harris' existing semiconductor businesses, which were then spun-off in 1999 as an independent company, under the Intersil name. In November 1998, Harris sold its commercial and standard military logic (semiconductor) product lines to Texas Instruments which included the HC/HCT, CD4000, AC/ACT, and FCT product families. Harris retained production of the Radiation Hardened versions of these products."

Spin-Offs

Good employees at an entrepreneurial organization like Radiation are themselves entrepreneurial, so it small wonder that many of them want to split off and do it on their own.

One of the first to do so was Parker Painter, director of the Orlando Research Division, who formed Dynatronics Inc. in 1957. He was probably motivated by a desire to stay in Orlando.

Another early offshoot was Systems Engineering Laboratories (SEL), founded in Fort Lauderdale, Florida, in 1961 by nine engineers from Radiation, Inc., Original founders were William Dodgson Jr., A.G. Randolph, Jack Jackson, Thomas J. Sullivan, Perry Knight, Edward H. Claggett, Larry Klingler, David Yoder, and Buddy Mock on January 21, 1961. They started in a warehouse with about 4,000 sq.ft of space at 4066 NE 5th Ave, in Fort Lauderdale. This location is considered to be the birthplace of the computer industry in Florida. Their initial interest was data acquisition systems using the technology developed for the Radiplex Low Level Multiplexer, but they soon were in the computer business. The company's first computer, the 820, was created as a special project for one customer, followed by the 810A, the 8500, and the 8600. In 1976, the Model 32/55 computer was introduced This system was the industry's first true 32-bit super minicomputer. The bus speed was 26.6 megabytes per second, which was a record at the time of its introduction.

SEL was purchased by Gould Electronics in 1981.

Chapter 10

Memorable Customers

A few customers stand out in my memory, some based on location, some from the individuals involved, some from the personality of the organization.

RADC

Rome Air Development Center, Rome New York stood out for several reasons. In the winter season, the climate alone was impressive, at least to a southern boy. The area gets an immense amount of snow, and it doesn't melt for months. They are good at removing snow from the streets, but there are often piles higher than your head along the roadsides. A good storm could close the highway for days, and I always feared getting snow-bound in Rome.

In contrast, the people were warm and friendly--they were mostly on Italian descent. The Italian food would spoil you for any Italian restaurants in other parts of the US. Our favorite Roman restaurant was always full of locals--many with a distinct Mafia appearance.

Our favorite motel was the Paul Revere, a local spot on the grounds of an old estate at the edge of town. The ambiance was not deluxe, but the rural location on the edge of a pond was a welcome change from our usual chain motel lodging on business trips.

Thiokol

Thiokol designed and built solid rocket motors for Minuteman in Utah, and purchased many data acquisition systems from us. Most of the Thiolol customers were Mormons and did not drink. This required a major adjustment in our normal customer entertainment practices.

NSA

Secure communications was a major part of our business and the National Security Agency was a major player. NSA was responsible for creating and breaking codes, but also worked in related areas such as voice processing and the interception of "bad guy" radio traffic. NSA was in a huge building at Fort Meade, MD, packed with computers. There were no (or precious few) visitor parking spaces, so you had to be prepared to search the vast parking lot for a space, which was invariably at the far corner of the lot--a long hike from the door. There were strict and rigid security procedures. A regular security form had to be sent in advance, but you also had to hand carry a copy, a step not normally required for other installations.

One NSA security task that is impressive in retrospect was the production and distribution of security keys to the entire government secure communications community. The keys were changed daily and were prepared in the form of paper tape which was physically carried to every comsec device in the network, a vast number, spread throughout the world. This cumbersome procedure was necessary because of the huge and on-going risk of transmitting them electronically. Today, public key encryption systems allow one to transmit secure traffic without any concern for key distribution.

NSA was slow and reluctant in implementing Public Key Encryption, whether from a "not Invented Here" syndrome, or the concern for the difficulty in "reading" a large volume of traffic generated by an easy-to-implement security system, or for some other reason.

This illustrates the paranoia-inducing contradictions that infuse the comsec field. A common goal is to make your security dependent only on the keys you use, not on the hardware used for encryption and decryption. (A vital contributor to allied code-breaking in World War II was the availability of captured

Axis coding hardware.) However, you still have to classify and protect the hardware--otherwise the enemy will get it and use your hard-to-break codes against you.

A related difficulty is comsec for friendly parties, even those within your own government. You would like for these to be good, but not too good--you might need to read them someday. If you indorse their use, people will think you have built in a secret "trapdoor" to allow you to easily read them. On the other hand, if you do not endorse them, people may be reluctant to use them. It may be hard to implement a suitably subtle trapdoor: should you plant a rumor that you have one anyway? Or will your most sophisticated enemy see right through that?

It may be that someone has recognized that classifying and encrypting your most important information is counter-productive and only flags it for priority cracking efforts. Maybe all our classified traffic is random data (making it impossible to break). Maybe all important traffic is hidden in the flood of tweets just with innocuous phrasing. Did you hear the rumor that the explosion of personal network traffic was secretly encouraged by government subsidies for just this reason?

MFM

Another important customer in the intelligence community will go un-named. One of their security rules was that their name would never be acknowledged in any connection with the company. If asked, we could only remark that it was a "proprietary customer." This is a wise rule; nothing is to be gained by publicity on the subject. It did generate the need for an internal code name, and that is an interesting story. When our security department was first visited by a customer representative, he was such a strange individual that he was referred to as the Man from Mars--hence the customer was named MFM, and so he shall remain.

A striking characteristic of this customer was that every individual we dealt with was an interesting person--someone you enjoyed having dinner or a beer with. This was certainly not true of every organization we dealt with.

Another striking characteristic of MFM was that they were excellent at managing contracts. Unlike many other customers they visited on a regular basis from the beginning of the contract, understanding what we were doing, and offering constructive guidance to better satisfy their needs. It was amazing the value of timely guidance to prevent us from inadvertently wandering from the customer's desires until it was expensive in time and schedule to correct.

On one of his many visits, my MFM customer noticed an rustic bar in West Melbourne, and expressed a desire to visit it.

"I don't think that is a good idea, Lou. Only local red-necks go there, " I said.

He insisted, and one of his areas of professional expertise was doing business in strange bars around the world, so I finally agreed.

We dressed down for the visit and pushed open the door. The room was dark, but crowded. It fell silent and we could sense that everyone in the room was staring at us. We advanced to the bar and were served promptly enough, but the room remained silent except for low whispers. It was clear that everyone but Lou and I were regulars, and that they were not often visited by outsiders. The beer was cold and tasty, but the ambiance was not warm. Lou was normally full of self confidence, but the strained silence of the room affected even him. When our beers were drained, he nodded toward the door and we paid up and left to find another bar.

FRB

Another interesting and quite different customer was the Federal Reserve Board. One of our technology developments was a high quality secure voice system. Our marketing people turned up a potential application. One of the routine procedures of the FRB was a periodic "Open Market" call between the Chairman and the president one of one of the district Federal Reserve Banks. The purpose of the call was to set the basic interest rate for interbank loans, which affected other interest rate around the country. Clearly, advance notice of changes in the interest rates could allow large illicit profits. Someone in government had noticed that the Open Market Calls, in the clear, had a risky potential for such a leak. The FRB had evaluated existing systems for securing the calls and decided they were unsatisfactory, considering the very senior participants, and the involvement of remote and varying cities around the country. We felt our system could overcome these objections and satisfy the bankers.

After much lobbying was were granted the privilege of demonstrating the system on a series of live Open Market calls. This required the installation of a terminal at the Chairman's office in Washington, and a roving terminal at the designated District office. I made the site survey visit to the Chairman"s office. The building was the most impressive government structure I ever visited. The halls were long and marble lined, with a huge version of the US seal on the floor at one juncture. The whole atmosphere was quiet and church-like. Clearly bankers want you to be impressed with their importance. The virtue of this to us was that the office suites were large and there was always a closet where we could tuck away our equipment. The district offices were a bigger challenge because we had to move our terminal in and install it in a limited time and without disturbing the operation of the office, and then move it out again in a few weeks.

I did the installation at several of the District Offices. I particularly remember Chicago because I had never been there before. I am not a big city person, but Chicago favorably impressed me, even in the snowy winter. The New York session was less fun.

The financial district in New York, where the Fed is located, is a spooky section of town. My hotel was on the fringe of it, and I could watch the sidewalks from my room. During daylight things looked normal, with crowds of visitors and office workers. As darkness fell, things changed. The daylight pedestrians disappeared and the sidewalks became deserted. Soon, out of nooks and crannies, appeared a few roughly dressed, spooky street people, who apparently controlled the area after dark.

I had planned to do my installation after business hours. When I tried to arrange a taxi to carry me and my boxes of equipment to the Bank, I found that most taxis were fearful of going in the area after hours. I finally found one to take the risk in the twilight.

When I finished the installation, it was late evening. The FRB guards had great difficulty getting a taxi to pick me up, and and finally one agreed to came only if the guard would meet him on the street. When the taxi arrived, the guard was reluctant to unlock the doors and actually drew his gun to escort me across the sidewalk to the cab when it arrived.

When we later removed the terminal, the boxes had to be stored overnight at the bank--in the basemnt vault containing the largest known monetary gold depository in the world--currently about 6700 tons of gold. If only we could have emptied our boxes and filled them with gold...!

NRO

The National Reconnaissance Office was an important customer, often in the role of a customer's customer in the complex world of satellite reconnaissance systems. I never had much contact with NRO until our Project 1300 was declassified, and I researched the background. What surprised me was that while NRO was operating in the blackest of black worlds, they had the foresight to hire an archivist to document the complex goings-on for future reference and historical use. And he produced interesting, readable reports. A classy customer!

WADC

Wright Air Development Center, at Wright Field, was an important early customer, responsible more that any other custormer for giving Radiation a toe hold in airborne PCM.

SATCOM

The US Army Satellite Communications Agency, located at Fort Monmouth, NJ was an importanct customer for modems, antenna and complete satellite communications terminals. The base was closed in 2011, with remnants moved to Aberden Proving Ground and Ohio. Fort Monmouth was usually reached through the New York city airports. The route down was mostly on limited access roads which were not friendly to missed exits. Dan McRae and I often were engrossed in conversations when visiting and missed critical turns, which required complex manuevers for recovery.

A TSC 54 Terminal is on public display in the Fort Monmouth area, as is a TLM 18 antenna.

Chapter 11

Timeline

The following table depicts a timeline of events over the period that radiation existed. Most of the information is derived from collections of Radiation Ink. For context, some notable events of the period, such as technological achievements, are included.

Year	Event
1950	Radiation founded
1950	First major contract—Timing System for AFMTC
1950	Fledgling Radiation employs 22 people in 15,000 square foot building
1951	First hardware delivered to Eniwatok atol
1953	Orlando Production Division formed
1953	First magnetic core memory in Whirlwind Computer at Harvard
1954	Orlando Instrumentation Division founded
1954	Contract for Operation Teapot awarded. (JA 1023)
1955	AKT14/UKR1, story on
1955	Fixed styli recorders, story on
1955	Malabar site begins operations
1955	First Radiation Ink published
1956	First stock issued (at $5 a share)
1956	Aero Commander a/c purchased
1956	Plant 7 occupied
1956	New Orlando Instrumentation plant opened
1957	TDMS license obtained from Automatic Telephone and Electric
1957	Orlando Research Division founded
1958	Sales $10M, Net income $488K
1958	Project Score, first US communications satellite announced
1958	"80 Acres" buildings occupied
1958	Plant 14 occupied
1958	Radiquad announced
1958	60 foot antenna installed at Kaena Point, Hawaii.
1958	First launch of Explorer and Vanguard satellites
1958	First working IC demonostrated by TI
1959	Sales $14M, Net income $588K
1959	Radicon/Radiplex developed
1959	First Palm Bay plants occupied
1959	Boeing Minuteman contracts received
1959	First Thiokol data system delivered
1959	Big Dish installed at Vandenberg AFB
1959	Project 1233 feeds designed in 77 and 91 days for 60 foot antenna in Hawaii and 250 foot antenna in England
1959	Doppler tornado radar delivered
1959	Contract from GE (Edwards AFB) for Data system
1959	Pershing missile contract awarded by Martin to Instrumentation Div (Orlando)
1959	TDMS contract awarded by Western Union
1959	Contract awarded by US Army for three Project Courier antennas

1959	Space Communications Division established in California
1959	Data Processing System delivered to Lockheed on JA6195
1959	Levinthal merged into Radiation
1959	Quick Look System delivered to Thiokol
1959	TDMS orders received from USAF
1959	Order for PCM Telemetry System received from Norair
1959	Orders received for 3115 Transmitters from Redstone Arsenal, Northrup, NRL, Chrysler, Hughes Aircraft and Motorola
1959	Last Vanguard launch
1960	Project Courier
1960	First ever low level multiplexer delivered to Thiokol
1960	First Bullpup missile trainer delivered to Martin
1960	Contract awarded from AC Sparkplug for PCM Telemetry for Titan ICBM
1960	First Minuteman PCM test set delivered to Boeing on JA 1297
1960	Data Processing System delivered to GE-Edwards AFB for J79 and J85 jet engine testing
1960	Levinthal awarded contract for radar transmitter for Arecibo, Puerto Rico.
1960	Tiros, first low altitude weather satellite, launched carrying two Radiation 3115 transmitters
1960	Products Division launched
1961	Contract awarded by RADC for microwave "space beacons" to study propagation through ionosphere
1961	Contract awarded by Bell Labs for tracking antennas for communications satellite program
1961	WATS telephone service installed
1961	First Data Systems for Nike Zeus delivered to Bell Labs.
1961	Contract for Nimbus telemetry announced
1961	New Aero Commander replaces older version
1961	Space Communications Division renamed Data and Display Systems
1961	Minuteman PCM (JA1290) shipped
1961	Radiation At Stanford receives contract for 100,000 watt transmitter for USAF
1961	Contract received for OAO PCM telemetry
1961	PCM telemetry delivered to Bell Labs for communications satellite
1961	SEL (Systems Engineering Labs) founded in Fort Lauderdale by Radiation breakaways including Bill Dodgson and Perry Knight
1961	Sales $26M, Net Income $587K
1962	High speed printer delivered to Lawrence Labs.
1962	In house Microelectronics (Integrated circuit) capability launched
1962	Dr Boyd hired
1962	Telstar antenna tracks Transit 4B satellite
1962	Two more PCM systems for Telstar shipped
1962	Minuteman Combat Training Launching Instrumentation system shipped to Boeing
1962	Paper published on use of chaff cloud for communications
1962	Minuteman launched with Radiation PCM system
1962	Telscom antenna demonstrated
1962	IRE merges with AIEE
1962	All TWX machines in Bell system switched from manual to dial switching
1962	Contract received from NASA-Marshall for PCM telemetry study
1962	Orlando operations merged with Melbourne divisions
1962	Radiation stock listed on AMEX, at $9
1962	Radiation donates $30,000 to BEC (now FIT)
1962	Data Logger for Ford shown
1962	Record low temperature of 24F in Melbourne
1962	Sales $24 M, Net Income $272K
1963	PCM systems for OAO delivered to Grumman

1963	85 foot antenna delivered to AMR
1963	Astronauts Stafford and Conrad visit Palm Bay plant
1963	Boyd elected President and CEO
1963	Proposal submitted to Intertechnique for PCM system
1963	Qualification Tests completed on OAO-SDHE and OAO-EDHE
1963	Radiation At Stanford 10,000 watt transmitter used to send TV over the Relay satellite
1963	Work started on PCM telemetry for Collins Radio for Apollo project
1963	National Airlines starts jet service from Melbourne to California
1963	Last Minuteman PCM systems on first contract shipped
1963	High Speed (30,000 lpm) Printer demonstrated for press
1963	Last OAO PCM equipment completed
1963	Radiation At Stanford sold
1963	Orlando manufacturing Division founded
1963	Sales $31 M, Net Income $670 K,
1663	Employee Headcount 2014
1964	Stonehouse operational
1964	Radiation 85 foot antenna at Point Mugu used for Olympic TV broadcasts
1964	Apollo and Saturn II data handing systems delivered to NAA
1964	First unmanned Gemini satellite orbited, using Radiation command system
1964	Contract announced for LEM PCM system for Grumman
1964	Aerospace Building opened in Palm Bay
1964	Apollo Data Processing System shipped to NAA
1964	Sales $44 M, Net Income $1.2 M,
1965	Radiation 12 foot tracking antenna on USAF ship track Mariner 4 launch
1965	First manned Gemini satellite orbited, using Radiation command system ground systems
1965	Contract awarded by NASA-Goddard for PCM-DHE for unmanned satellite tracking network
1965	IBM 360-30 installed in Palm Bay, replacing IBM 1620
1965	Programmable Diode Matrix from Physical Electronics Division displayed at IEEE show.
1965	Contract awarded by USA for nine Mark V satellite communications terminals
1965	DAS-10 Distortion Analyzer System announced
1965	First AMARS delivered in 30 days
1965	Lear Jet purchased
1965	Contract awarded for Concorde recording system
1965	Contract awarded for AN/ASW-25 communications system
1965	Contract awarded for advance Minuteman system from Boeing
1965	Radiation purchases Paricode and Locotrol supervisory control product lines from North Electric Co.
1965	Contract from NASA announced for IRLS
1965	Simulation CRT Display System for FAA announced
1965	Products Division and Control Systems Division merged to form Control/Communications Division
1965	Orders announced for Locotrol and DAS-10 units
1965	Radiation gives $20,000 to BEC
1965	PDP-8 Minicomputer introduced by Digital Equipment Corp (DEC)
1965	Sales $38 M, Net Income $1.2 M 2506
1965	Employee headcount 2506
1966	First OAO orbited, carrying Radiation equipment
1966	Control and Communications Division delivers first product-based data acquisition system
1966	Physical Electronic Department becomes Microelectronic Division
1966	ASW-25 passes first flight tests aboard USS America
1966	Sales $46 M, Net Income $1 M
1966	Employee headcount 2768

1967	Employee headcount reaches 3000 in April
1967	Merger with Harris
1967	First dual-control Locotrol ordered by N&W RR
1967	Palm Bay building designations changed from letters to numbers
1967	Radiation is prime contractor for IRLS (Interrogation, Recording and Location System)
1967	Significant MED orders from Raytheon, Sandia and Lockheed
1967	Major Programs listed: Interim/Advanced Defense Satellite programs for USA, ASW-25, IRLS, Nimbus satellite and ground equipment, Advanced Minuteman, Concorde, Apollo and Saturn equipment, Lunar Module, OAO, Titan III, ALTAIR, Lunar Orbiter
1967	CENTRIX telephone system installed
1967	Advanced IRLS contract awarded by NASA
1967	Fire in Building 6
1967	IBM 360-50 replaces IBM360-30 computer
1967	Contract awarded for 60 foot antenna for UK.
1967	First Hexagon contract awarded to Perkin Elmer by government
1968	RA-909 Op Amp announced by MED
1968	Contract awarded for PCM for USAF Manned Orbiting Lab.
1968	Space Ground Link System (SGLS) modification completed in 90 days
1968	Tullis elected CEO
1968	Contract received from Grumman for Data Link for A6A
1968	VAST program announced
1968	Employment reaches 4,000
1968	60 foot VHF, UHF, S-and antenna installed at Keana Point, Hawaii
1968	RACE demonstrated
1968	Employee headcount 4027
1969	Apollo moon landing
1969	Harris merges with RF Communications
1969	MED displays radiation hardened circuits at IEEE show
1969	Controls Division ships CAPLINE supervisory control system to Shell.
1969	50 Locotrol systems ordered from Controls Division
1969	Deilguide antenna modification developed for AN/FPS 16 tracking radar
1969	Add-on contract won for JA1526 Flight Line Test Sats
1969	Dual antenna system developed for aircraft carrier satellite communications
1969	Britain's Skynet, using Radiation 60 foot antenna, becomes operational
1969	60 foot inflatable radome installed near Building 6 for antenna tests.
1969	Contract awarded for data Link for A-7E
1969	First use of Arpanet, predecessor of the Internet
1969	First software word processors marketed by industry. (Did not use CRT displays)
1970	Elk tracking experiment using IRLS described
1970	Model 1100 Display Terminals, 2500 Newsroom Systems, 1500 Video Typewriters and 2200 Video Layout Systems introduced (commercial CRT word processors did not appear until 1972)
1970	Harris 1100 Editing and Proofing Display demonstrated
1970	Control Division awarded several contracts for supervisory control systems
1970	Totty named ATD Director
1971	Hexagon launched
1971	2415 ASW-25's have been delivered
1971	Two 60 foot S-band antennas ordered for Wallops Island SMS ground station
1971	2250 pound LaFaire Vite proposal shipped
1971	Double mesh deployable antenna shown
1971	Northern Radio (subsidiary of RF Com) moves to Melbourne
1971	Harris 1100 Editing Terminals sold in Europe

1971 Disney World opens
1971 Signal processors for Earth Resources Technology Satellites (ERTS) shipped
1971 Low cost satcom antennas developed
1971 Damaged Shipboard antenna on USS Constellation replaced in 25 days
1971 First VAST system shipped by PRD
1971 First Hexagon launch
1972 "Video Typewriter" terminals for Florida Today announced.
1972 VAST celebrates 5th year
1972 Composition Group announced to build publishing terminals
1972 Advertising terminal announced for display advertising
1972 Frequency selective feed for UK-TCS
1972 Controls Division becomes Harris Controls
1972 W-71 for Aerospace Electronic Systems Company becomes division's largest program
1972 Orders for more Locotrol units announced
1972 1300 (Hexagon) becomes largest program
1972 78 foot diameter radome constructed to test space deployable antennas
1972 SETAD aerospace program for Navy described
1972 Contract for DCL (Direct Communications Link) between US and Russia announced
1972 New S-band feed installed aboard USNS Vanguard tracking ship
1972 Contract awarded for DCL, a hotline connecting Washington and Moscow
1973 Radiation name dropped
1973 Emux System for B1 bomber announced
1974 Contract for three weather data stations awarded for 5D/GSM program
1974 Upgrades announced for TSC54 Satellite Communications Terminals first built in 1967
1974 Two Modems built for Skynet in 10 days
1975 Radiation Ink is now Harris Ink

Chapter 12
The Community

The South Brevard area was quite a different place in the 1950's and 1960's when most old Radiators first came to town.

Melbourne and Eau Gallie were separate towns and had a distinct gap between them. The total population was only a few thousand. Palm Bay hardly existed. The only beachside communities were Indialantic and Melbourne Beach. Highway A1A between Indialantic and Patrick AFB had virtually no buildings--only palmettos. South toward Sebastian Inlet, A1A was newly paved: Honest John's was still referred to as at "the end of the old hard road."

The buildings of downtown Melbourne are virtually unchanged, although the tenants are different. There was McCrory's "ten cent store" which carried all sorts of useful items. There was at least one drug store, Taylor Drugs, complete with a lunch counter. At some point, there was also a Fulmer's Drugs.. Hatt's Bait and Tackle was on Melbourne Court, next to the railroad tracks. It included Pool and Snooker tables. Shrimp were 3 dozen for one dollar. There was a Coleman's Bakery on New Haven. Shorty's Bar was on New Haven and the Melbourne Bar on Melbourne Court.

The downtown section of Eau Gallie was much less developed. The impressive bank building at the corner of Highland Ave and Eau Gallie Blvd is most notable. Other old downtown Eau Gallie businesses included Karrick's Grocery, Mather's Bakery, and Coleman's Drug Store (unrelated to Coleman's Department Store in Melbourne, but operated by a brother in Coleman's Drugs in South Melbourne). On US1 at Ballard Drive stood Jackson's "Sooper" Market (It may have had other names over the years). It was a two story affair and carried some hardware and miscellaneous items. The Boathouse presently occupies this building.

Housing Sub-divisions

Many Radiators lived in Magnolia Manor, conveniently located near the airport. Dan McRae, Earl Smith, and Scott Campbell come to mind. (Located within a one or two block radius on Norman Drive and Bonnie Circle were Mel Cox, Harvey Bush, Earl Damaan, Gantt Hamner, Bob Dryden, Julia Ruhlander, A.B. Amis, Guy Pelchat, Bu Chen, Marge Gilbert, Jessie Ferguson, Teddy Van Eeghan, Ed Dorsett, and Harold Caudill.) Old Loveridge was in Eau Gallie, east of US 1 and south of the Eau Gallie River, and that's where Radiation big shots Homer Denius, George Shaw, and Bill Dodgson lived. Frank Perkins lived in a rental house there when he first moved to the area. New Loveridge was west of US 1 in the same area. Jim Coapman lived there, as did John Spooner, Parker Painter, and Ralph Johnson. (Ralph later moved to a waterfront house on the Eau Gallie harbor.) Almar, was a newer subdivision in the same area. Frank Perkins owned a house there, as did Cy Underwood, Robey Green, and later Harvey Bush. Between Babcock and US 1 east of the airport was a popular residential section known as Indian River Bluff. Hans Scharla-Nielsen had a garage apartment there. Melbourne Village was (and is) an interesting residential area. Perry Knight, Cole Goatley, George Hedman, and Frank Perkins lived there at one time. Ernie Ploeger still does.

Shopping

No malls or shopping centers in the 1950's, but by the mid 1960's the Melbourne Shopping Center had been constructed on the east side Babcock, and Brevard Mall on the west side. Lovetts, on the south side of New Haven where a same day surgery center is now located, was one of the few grocery stores in town and you were out of luck if you needed anything after the 5:00 pm. closing time. Any serious purchase required a trip to Orlando. There were two ways to get there, neither ideal. The Kissimmee Highway (US 192) was full of bumps and undulations. The sharp, bumpy curves near the St Johns River bridge were justifiably known as Dead Man's Curve. A priest and two nuns died in one accident there. The other route involved US 1 to Cocoa, which had heavy traffic on two lanes. The normal route from there was up to SR50 at Indian River City and across. A short cut was just opening, the Bithlo Cutoff, now SR 520.

Restaurants

Thee were a number of good restaurants. The Sea Room (formerly Kokomo Drive In), on US 1 near Babcock, had good affordable seafood and a warm welcome by the manager "Tommy" (Helen?) Jorgensen. (There was a Jorgensen's Fish Market on Crane Creek, and a Jorgensen family in Grant. The Jorgensen's General Store is a historic site in Grant-Valkaria, Florida. It is located at 5390 U.S. 1. On June 25, 1999, it was added to the U.S. National Register of Historic Places, NRIS #99000711.) Henchel's Red Rooster was another area restaurant. Marshall Richard's Melbourne Beach Steak House was located on A1A just north of Ocean Ave -- there's a library on that location now. Great steaks, hot twice-baked potatoes and Gomez dressing on your salad, and service by long-time waitress Joan. In downtown Melbourne was Pop's Casbah. The name is still on the wall at Matt's Casbah, but the place is different. The old place had only a verbal menu, but a long one fired at you at top speed. You didn't dare ask for a repeat. Other people remember Pop's as serving very good peanut butter pie. Another feature was very low ceiling in part of the restaurant. Another downtown cafe was Helen and John's Grill, on Strawbridge. Ron Winn has an undated menu from there listing a Cube Steak dinner for 90 cents. Another Melbourne spot was the Ever Ready Dining Car, on the East side of US 1 across from the Van Croix, in an old train car. It may have been gone by the 1960's. In Eau Gallie on US 1, and perhaps at a little later in time was Tippy's Taco House. It was a favorite for engineering lunches. The printed menu listed lunches lettered B, C, D, etc., but didn't have an "A", for some reason. Try the "wet burrito". The original local McDonalds, on Hibiscus across from Montgomery Wards, is still doing business at that location. Before McDonalds, there was the Burger King, in its present location near the ice plant. They had a hamburger machine that grilled hamburgers and dropped them in a pan of sauce. Lum's was on Babcock St. just north of the present Hooter's.. The building is still there. There was a cement plant across the street. Another popular lunch spot was a weird cave-like stucco place in West Melbourne, the Camelot, I think was the name. They had really good pickles. Pete's Italian Restaurant out west on New Haven was a favorite lunch spot of Radiators from RF Division, and a favorite sub there was the "Manweiller Special" named after Harlan Manweiller from Soroban. Pete's later located to Satellite Beach adjacent to the Satellite movie theater and operated under the name "Anthony's". A couple of daughters formerly from Pete's later ran a small sandwich shop named "Rosie's" in a Dairy Queen building on US-1 just south of Turkey Creek in Palm Bay-- great Philly steak subs with peppers, onions, and cheese (and a dash of Tabasco to liven things up), eaten outside under towering live oak trees overlooking the river. Another occasional lunch spot was the Cox's Truck Stop on US 1 in Palm Bay. Jerry's Pizza Parlor at the north end of Babcock. had good Italian food. For a splurge a little further afield, there was Bernard's Surf with all sorts of weird stuff on the menu (like chocolate covered bees), Another Cocoa Beach splurge was Ramon's, the "Toast of the Coast." At any of the Cocoa Beach spots you might see one of the original Mercury Astronauts. A unique spot was Miguels Posada del Rey on New Haven near Babcock, where the CVS Pharmacy is now. The structure had an interesting history. It was built in 1926, for William "Doc" Sloan, a well-known local bootlegger. Shortly after he moved in, their 2 year old daughter Cora was playing with matches and died of her burns. Her spirit was said to haunt the house. Sloan later went to prison for his bootlegging activities, and the family moved to Fort Pierce. The house was used by navy personnel during the war, and later was used as clinic and hospital by a Dr Pennington. The Mexican food at Miguel's was good, but the Margaritas were outstanding--large, strong and tasty. It is a shame that such a historic building was replaced by a chain drug store!

At A1A and Eau Gallie Boulevard was the Pelican Restaurant, featuring Rosin Baked Potatoes.

A couple of places on Palm Bay Road have been around for some time--I am not sure how long. The Pepper Tree serves fresh-made sandwiches. Rooney's has been in business for over 25 years, and had several names before that, including "Town Pump."

One of the classiest restaurants in town was Poor Richards (now Djon's) in an old house in Melbourne Beach.

Mrs. Edge's cooking was legendary. First she had a large restaurant just west of Harry Goode's Outdoor Shop. Then she "retired" and opened the "Town House" across Waverly from Pop's

Nissen's restaurant was located where Paisano's is located now. Bernie and his mother were hams. If you ate there, they would get you a "phone patch" to anywhere.

Bill Sargent writes:

"The Riverview Restaurant on the SE side of Melbourne Harbor was a landmark into the 1950's. It was a two story building with the kitchen on the ground floor and the dining room on the second floor. It had a dumbwaiter that took the food to the second floor. The building is gone now.

Bars

There were a number of notable bars in the area. In downtown Melbourne were Shorty's Bar on New Haven and the Melbourne Bar on Melbourne Court. Both were dark and oriented to "locals". Shorty's included a billiard parlor where mobster Al Capone is reputed to have played. Rick's Oak Tree was located under a big oak on Babcock. Smitty's bar was located further north on Babcock. West on New Haven was the Moonlight Tavern where I once took a customer, at his request. (The visit was short.) The Dixie Bar on west New Haven Ave. was a popular businessman's spot. There was also a place called the Library in West Melbourne. In Palm Bay, on the Indian River, was The Castaways Club, surrounded by lush plant and palms. On the beachside, Dragon Lady's Den was on A1A just south of the Eau Gallie causeway. The building still houses a bar.

The Blue Front Inn was on the east side of US 1 opposite the Van Croix theater, in the 950's.

Grocery Stores

Grocery stores were fairly rare and were open shorter hours than we are accustomed to now. There were an A&P, Winn-Lovetts' and an IGA on New Haven Ave. in downtown Melbourne, and Bob Kempfer had one in the same area. Ed Kempfer had a store on the South Dixie Highway. In Eau Gallie, there was a Karricks Grocery.

Drug and Variety Stores

Downtown Melbourne had several drugstores: Fulmer's and Taylor Drugs. Coleman's Drug Store was in south Melbourne. The area also included a McCroy' variety store. Hendrick's was a popular soda fountain and an office supply/variety store which eventually morphed into present day Meehan's. In downtown Eau Gallie was another Coleman's Drug Store. On US 1 at Ballard Drive, Jackson's Supermarket sold hardware and miscellaneous items.

Department and men's wear stores

Department stores were concentrated in downtown Melbourne. There were two department stores on New Haven Ave.: Turner's (which carried official Boy Scout clothing and equipment) and Coleman's. Turner's was opened in 1945 by John Turner, who arrived in town in 1937, and was principal of Melbourne High School, then on New Haven Ave in what is now the Henniger complex. There were also two men's wear stores, Dennis-Medvine, on Melbourne Ave. and Kemper-Jennings on New Haven.

Bakeries

Hedges' Bakery and Coleman's Bakery were located in downtown Melbourne. There was another Coleman's Bakery in Eau Gallie, as well as Mather's Bakery.

Hardware Stores

Huggins Supply was a large hardware store east of the railroad tracks. Painted on the wall was: "Melbourne's first phone number, 1." Another hardware store in downtown Melbourne was Delisles. Broom's, was located on South Dixie Highway next to Coleman's Drugs, both in downtown Melbourne. There was an Eau Gallie Hardware in Eau Gallie, where Ace Hardware is now located.

Auto Parts and Repairs

Two old businesses in this category thrive to this day. Cassels Garage has been in business on New Haven for a long time and its history is intertwined with nearby Ferguson's. Howard Fulton, the present president of Cassels, writes: "Lester Cassels (my Dad) came to Brevard in 1950 by the request of Hollis Ferguson & Carl Schmidt. Hollis was Service Manager & Carl was General Manager of Fordyce Chevrolet in Melbourne. Fordyce Chevrolet was just south of Crane Creek on US1. Later the building became Mac's Friendly Tavern. Dad worked at Fordyce Chevy for a couple of years before being drafted into the Army. Dad served his country until about 1954 when he came back to town. Upon arriving, he went to work for Hollis Ferguson at his new store in West Melbourne. Dad ran the service department & Hollis ran the parts department. Lester Cassels met & married my Mother in 1964 . I was 6, had a younger brother 6 months & three older sisters 8, 10 & 12. Marrying Mom & raising 5 children was, I think, by far Dad's greatest accomplishment. Dad & Mom built and opened Cassels Garage in 1966. I swept floors, cleaned bathrooms, cleaned parts, and learned all about cars starting the summer before I entered 5th grade.

"I am very proud of where we came from and how we got to where we are. The business has survived many milestones, from wars to hurricanes, from LBJ to George W.

"I, my wife Bobbie, and my children Jessica and Katie are carrying on the family business proudly - you'll notice even the phone number is the same 40-plus years later! We anticipate serving the community honestly & fairly for many years to come."

The waiting room at Cassles holds a financial notebook from 1966, which contains a lot of interesting company and local names as customers. Take a look at it when you are in the shop.

Ferguson's now concentrates on trailer parts and repairs, and are the "go to" place for boat trailer repairs.

Bait and Tackle Shops

Harry Goode's Outdoor Shop has been in business since 1946. Their web site reads:

"Harry Goode Sr. was a descendent of Jesse and Richard W. Goode who were one of the founding families of Melbourne. He enjoyed fishing and worked in a fishing store before going to serve in World War II. After returning from the war, he started his own fishing business out of his garage, then relocated to the present site of 1231 E. New Haven Avenue in 1946. His wife, Katharine Goode, kept the books and did reports and orders. Harry rented half of the building while Brooks Paint and Glass occupied the other half. When Brooks Paint and Glass relocated, Harry obtained their half and opened a Schwinn Bike Shop which was in business until 1993 when they decided to expand their fishing line.

Some of the original products sold at the Harry Goode's Outdoor Shop were guns, rifles, ammunition, 3 wheel bikes, cane poles, lanterns, knives, Shakespeare fishing rods and rigs, Pflueger fishing tackle and archery equipment.

Harry had some unique trademarks down at the store. He often weighed catches on a back scale, then he would trace the outline of the fish with chalk on the floor and sidewalks. Often fisherman would stop by the store and share their catch with him. Harry hosted a well known fishing contest and he gave away over $1000.00 in merchandise as prizes. He did a daily radio report which gave tide reports as well as what was running and where the best spots to fish were. His motto was "fish more, live longer"!

Harry Goode retired in the 70's but still visited the store. He spent summers in Franklin, NC, wrote a book, gardened, painted and enjoyed life. His sons, Harry C. Goode Jr. and Richard W. Goode Sr. took over the store. In 1988 Harry's grandson, Richard W. Goode Jr. came to work at the store.

Harry Goode's Outdoor Shop continues to be a family business that provides integrity, superior customer service, product availability and loyalty to its customers the way that Harry did back in 1946."

Harry had a huge gun collection. He also had a giant rattlesnake pelt over the door to the back room.

For years a colonial era canon rested in front of the store--appropriate since the C in Harry's middle name stood for "canon." The gun now resides on the grounds of a restaurant on the north side of New Haven.

From their Facebook page, Hatt's reports:

"In the early 1940s, Russel Hatt Sr. opened a bait and tackle shop located where what is now an antique shop, next to the railroad tracks, in Historical Downtown Melbourne. When his son Jerry Hatt, a self-taught diver, bought an air compressor from a military surplus store in Orlando, he began to fill tanks for himself and his friends. Jerry soon took over a corner of his father's bait and tackle shop and began filling tanks and selling dive equipment. As Jerry got older he married and moved to what is now the Chart House located on the Melbourne Harbor. There Jerry operated a bait and tackle and dive shop from 1960-1974. In 1974 they built a brand new dive facility complete with dive training pool which is our current location. From the first bait and tackle shop to the current dive shop all the Hatt businesses have been within 1 block of the others.

"Current owner Mike Hatt, son of Jerry Hatt, can honestly say he has grown up in the dive business. From the day he was born Mike lived on the second story of the bait and tackle dive shop. As a young toddler to current day diving has been a prominent part of Mike's life. Mike married Starr Hatt and is currently operating the store with her. Their four children, fourth generation scuba divers, Crystal, Joshua, Kendra, and Jarod Hatt play an important role in the dive business. To us diving is not just a recreational sport, but a way of life."

Kline's Bait and Tackle was on the south Eau Gallie causeway just west of the Indian River. Live shrimp were $1 for three dozen. There was later a sea food market in this location.

For freshwater, there was Rotgers' bait shop on the north side of US 192 in West Melbourne. They sold earthworms and minnows.

Banks

The Bank of Melbourne was on New Haven near the railroad tracks. The Bank of Eau Gallie was in downtown Eau Gallie, in a "bank looking" building with big white posts. There was another bank west of the US 1 in Eau Gallie, but I don't remember the name.

The local banks were more tuned to fishing and agriculture than electronics and would not advance money for government contracts held by Radiation. So Homer Denius and a group of local businessmen started the First National Bank of Melbourne. The bank and Reverend Alex Boyer were responsible for starting the Trinity Tower development here for the elderly. The bank also helped finance FIT. The First National Bank of Melbourne was sold to the Sun Group after the merger of Radiation and Harris-Intertype, and the Sun Bank continues to be one of the principal banks in the area.

Hotels

The Melbourne Hotel opened in 1924 at the corner of New Haven Ave and US 1. The building today is known as the 1900 Building. This hotel opened on September 23, 1924, with an arcade going around the building, and a garden in the back. Flower shows and teas were held there during the tourist season. The Melbourne State Bank opened an office in the hotel in 1924. It once had a barbershop and a Western Union office opening off the lobby, and a Fulmer's Drug Store on the corner.

Sidney Platt began working in the hotel in 1925 and served as its manager from the early 1930s until 1969.

The Belcelona Hotel, which now houses the Florida Air Academy, had an interesting history. It was built in 1923, for Ernest Kouwen-Hoven as a home called Magnolia Manor. In 1926 it was remodeled into the Lincoln Hotel, and later renamed the Belcelona Hotel in 1957. The property was sold to the Florida Air Academy in 1961. Bill Eddins used a room there for teaching transistor classes around 1959.

In Eau Gallie, the Oleanders Hotel stood on the banks of the Indian River just north of the causeway. It was later called the Imperial Hotel. Squidlips restaurant now occupies the site and incorporates a few fragments of the hotel building. The original hotel on the site, the Harbor City Hotel was built here in 1925, with Dr. W.J. Creel president of the firm owning it. Opening ceremonies were held at the hotel May 8, 1926.

The Tradewinds Hotel began life as the Hotel Indialantic. It had its Grand Opening 1923 and featured a floor of Vermont marble, luxurious rooms, a grand lobby and a beautiful dining room. It also had a pool and golf course. . As a result of the depression it closed in 1932, 1933, and 1934. In 1935 it was purchased by Tom Doherty and reopened. The Doherty family lived there for several years. It featured a list of famous guests: Edward G. Robinson, Vera Evans, Jimmy Doolittle, Jack Benny, Zack Mosley, Frances Langford, Werner Von Braun, Charles Lindberg, Gen Matt Ridgeway, Edward R. Murrow, and Gov. Leroy Collins.

The Tradewinds hosted the Radiation Christmas party in 1959. One summer they offered memberships in the Olympic size swimming pool and lots of Radiators joined.

FIT owned it briefly and used it for dormitory in 1969. In the 1980's the property was bought by developers and the building was demolished in 1983.

The Midway Colony on south US 1 was an interesting lodging. It was a collection of small cabins with recreational facilities on the grounds like shuffleboard and a meeting room. It was primarily occupied by tourists on extended stays. It was named because it was mid way between Jacksonville and Miami. It began in the 1920's as a tent camp and at one time had it own power plant. The old site is now the park where the Radiation Old Timers reunions are held

I recall there being several motels along US 1 north of New Haven, but the only name I recall is the Goff Motel. It was previously called "Crane's Court."

Indialantic had several tourist motels along the beach. The present Tuckaway Shores facility was once called Sharrock Shores Resort.

Beach Clubs

The Bahama Beach Club in Indialantic was popular, with a swimming pool, restaurant and bar. It originally operated as the Indialantic Casino and was renamed in 1944.. I recall several dinners there as guest of Jim Coapman, who was my boss at Missileonics before I went to work for Radiation. The swimming pool was the big attraction, with a barely submerged wall separating the kiddies' wading pool from the larger pool.

The Melbourne Beach Casino was a recreational complex south of 5th Avenue in Melbourne Beach. It had a pair of sulfur water swimming pools and a restaurant and bar. In spite of the name, it never operated as a casino.

Doctors and Dentists

Both were in short supply in the early days. I remember Dr. Arthur Tedford, who had an office on the north east corner of Strawbridge and Waverly Place. The building is still there.

My long-time dentist was Dr. Dean Hambel, who had on office on Babcock in Eau Gallie (near where the bowling alley used to be.) Dr. Dean died in 1972 and his brother Dr. Bill took over the practice until he passed it on to Dr. Sangiv Patel in 1995. His current office is on US 1 in Eau Gallie. Another long-time dentist was Dr. Peters, who retired and sold his practice to Dr. Tim Lewis.

Dr. Ted Kaminski, and Dr. Gayden shared a practice just east of Meg O'Malley's present location.

TV

This was long before the time of cable TV. The nearest broadcast stations were in Orlando, and required a tall, precisely aimed antenna for even marginal reception. There were three channels: 2, 6, and 9. Channel 2, the NBC station, was normally weak but usable, but when propagation conditions were good, the signal was strong. Unfortunately, the good conditions meant that another, more distant station on the same bearing interfered strongly with the Orlando station. You couldn't win. CBS was on channel 6 and ABC on channel 9.

Local Radio

Brevard's first radio station was WMMB. It originally went on the air on May 8, 1948, on 1050 kHz, broadcasting from sunrise to sunset. It was part of the Liberty Broadcasting System, and had a power of 250 watts. In June, 1951 WMMB moved to 1240kHz and began broadcasting 24 hours a day. From 1949 until about 1952 Chick Catterton was General Manager of Melbourne Broadcasting Company, owner of WMMB. In 1956 he was Vice President, General Manager and News Director of WMMB. In 1958, he co-founded WMEG. Chick was also a musician, and founded the Melbourne Municipal Band in 1965. He was mayor of Melbourne in 1953 and 1954.

WMEG began broadcasting in 1958.

Movies

The only indoor theater in the area was the Van Croix, on US1 north of the present 1900 Building. The Van Croix was opened in 1925, and had seen better days by the 1950's. The balcony was reserved for the colored patrons. The building was demolished in 1974.

There were several drive-in theaters. The Melbourne Drive In was located off US192 just east of the present Melbourne Square Mall. Some of the Australian pines lining the entrance road are still there, visible from the east end of the mall parking lot. Australian pines were once common, planted mostly as wind breaks but these are some of the few of these trees remaining in the area. A killer freeze claimed most of them in later years. The Brevard Drive In Theater was at the north end of Babcock St near US 1. On Sundays, on a lot in front, Hugh Campen, nicknamed "Peanuts," sold rides on a small (maybe 1/8 scale) steam train he built and ran as a hobby. (Incidentally, Hugh Campen was also the projectionist at the Van Croix theater.) There was also a Beach Drive In, off South Patrick Drive in Indian Harbor Beach, but I don't remember much about it.

Yacht Clubs

The Melbourne Yacht Club is now located on Melbourne harbor just south of the Crane Creek bridge. Hasty Miller writes of its history:

"For many years, there was a marina on the east side of the Melbourne Harbor [near the present site of Ichabods}. It was owned in the 1930's by one of the Beaujeans, but was bought in about 1946 by Walter Masland, who needed some place to keep his 49-foot yawl "Anchorite". After "Mas" moved into Melbourne, he decided that we needed a yacht club, so he and several of the prominent people around town started one. [records indicate 73 charter members in 1947] Walt sold them the land immediately south of what was by then the Indian River Marine Basin. In the '30's, there had been a club called the Melbourne Sailing Club, so they tagged onto that, appropriating the club flag, and, I am sure, many of the old members. I joined in 1967, and a couple of years after I joined, they built the new docks. They had professionals put in the pilings and the collars (the transverse members), and we, as a club, put in the new stringers and decking. I was monumentally impressed with the organization of that project. We had four teams working out on the docks, nailing down the new boards. We had a team on land cutting the boards to length, and a team with small boats ferrying the boards out to the three teams that were working on the docks, but had no access to land since the dock wasn't decked yet. Everyone was working and no one was "supervising". I can't remember whether we got it done in one day, or one weekend, but it went awfully fast. Many years later, (1980?) we bought land from Jackson Vaughn where the club is located now."

Commodores from the 50's and 60's were: Hans Scharla-Nielsen, Steve Batchelor (Soroban), George Shaw, Dwight Moore, Arthur Tedford and Albert Tuttle.

The Eau Gallie Yacht Club was organized in 1907 and the first Commodore was George F. Paddison. Three years later George and Gertrude Paddison deeded the property on the east side of Houston Street in Eau Gallie, fronting on the Eau Gallie River, to the Club, where the first clubhouse was built and occupied until 1960.

The membership voted in 1959 to dispose of the old clubhouse property and seek a suitable location on which to build a new home and operate a full-time Club, with adequate docking and parking facilities. The old clubhouse was sold and is now a private residence. The Gleason family generously gave to the Club over three acres near Mathers Bridge on which the new clubhouse was built in 1961. The Club roster contained slightly over 70 names when the membership expansion program was started in 1959. This was increased to three hundred as of the opening of the new clubhouse on December 23, 1961. Since that time there has been a continuous program for facilities development. The property is now valued in the millions, and the current membership is approximately 900 families.

The "new" Eau Gallie Yacht Club was formed when Homer Dennius and Charlie West (President of Soroban) needed a nice place to take customers for a good, quiet business lunch. They would have built themselves a place but Florida said that they had issued enough liquor licenses and were not going to issue any more. EGYC already had a liquor license, so Homer and Charlie took over the old, nearly defunct club.

Piers

There were several ocean piers. Canova Pier was at the end of the Eau Gallie causeway, and was popular with fishermen. There was another pier at the old NCO club at Patrick AFB.

Roads and Bridges

Babcock Street was gravel north of the railroad tracks (remember the airport beacon light and Smitty's Bar at that junction?) until Melbourne City Commissioner Ed Dorsett got it paved, and we thereafter jokingly referred to it as the "Edward A. Dorsett Memorial Parkway". Wickham Road was a work in progress, built over time by county commissioner Joe Wickham as manpower and equipment were available. Sarno Road was unpaved west of the railroad tracks, but there were rumors of residential construction to come. A1A between Canova Beach and Patrick Air Force Base was two lanes, with virtually no buildings along this stretch. A1A hadn't yet been paved to the Inlet in the early 50s, and there was of course no bridge spanning the Inlet. The entrance to Honest Johns, and the old Coastguard Station, was still described as at "the end of the old hard road." A few determined fishermen owned beach buggies that could navigate the sand tracks leading from "the end of the hard road' on down to the Inlet, and a few had even built shanties near the north jetty. The US 1 bridge across the main channel of the Eau Gallie River was called the Humpback Bridge for its arched shape. It was a narrow two lanes, with rough concrete railings which took a toll of carelessly driven cars. You dreaded meeting a wide truck. The southernmost channel (Elbow Creek) was crossed by the "Toothpick Bridge", named for the rickety wooden pilings. The Eau Gallie Causeway bridge was wooden and very narrow (but great for fishing and shrimping because it wasn't very high above the water) and had a swing span for river traffic to pass through. It had barely enough room for two cars to pass, and it had an almost right angle turn toward the east end.

The first bridge (wooden) across the Indian River at Melbourne was completed in 1921. The low level concrete replacement was finished in 1947, and the present high rise finished in 1985.

Another bridge was Mathers Bridge across the Banana River at the southern tip of Merritt Island. The draw was opened and closed with a manual crank. It was a great place to fish: low to the water and not much traffic. At the Merritt Island end was a good place, maybe called Mather's Restaurant, for a hamburger and beer. Parking was a problem, however. This place often had Highwaymen paintings for sale. The Florida Highwaymen were a group of self-trained artists who sold their work to individuals and businesses from the mid fifties. Many of the inexpensive paintings have appreciated greatly in recent years.

Miscellaneous

Anderson's Dig

A recluse named A. T. Anderson dug for ancient bones far west of Malabar. He was known as "The Bone Digger" and operated a small museum out of his house. He claimed that some of the Ais Indians who had inhabited the area were 7 feet tall, and he had many bones to prove it. He also claimed to see and converse with "spirits" of dead Indians who had inhabited the area. He also trapped rattlesnakes and sold them live to Ross Allen at Silver Springs. Yale University conducted research at the site, as reported in the following references, which are out of print:

A Survey of Indian River Archeology, Florida. Irving Rouse. (296 pp., 8 pls., 15 figs., with appendices by C. D. HIGGS and C. W. Goer. $4.00 with No. 45. Yale University Publications in Anthropology. No. 44, New Haven, 1951.)

Chronology at South Indian Field, Florida. VERA MARIUS FERGUSON. (62 pp. 4 pls., 10 figs., with appendix by M. War WINKLE Hove. Ibid., No. 45. New Haven, 1951.)

Houser's Zoo

The zoo was located on the north side of US 192 in West Melbourne. Houser, a citrus farmer, started exhibiting a couple of rabbits and a kangaroo in 1960 to attract people to his orange groves. By 1985, his roadside exhibit featured 250 exotic birds and animals, including rare pileated gibbons (an Asian ape). The animals were eventually donated to the current Brevard Zoo.

I can recall once in the early days when an itinerant animal keeper brought an elephant by the zoo and offered rides on the animals.

Ice plant

The old ice plant on US 1 is on the National Registry of Historic Places. This two-story Masonry Vernacular style building was erected in 1926-27 to serve households, businesses and local fishermen. It had a 50-ton capacity and cost $150,000 to build out of a steel frame, tile block and stucco. In the early days, the phone number was "2." Russel Vann was the manager for years.

This was the most modern and one of the final ice plants constructed by Florida Power, which later found that it was more profitable to generate power than it was to produce ice. After a 1941 reorganization of Florida Power, this building was leased to City Ice and Fuel Co., a Chicago company which continued to produce ice in it until 1977. Ice was delivered to merchants in the downtown area

This was the most modern and one of the final ice plants constructed by Florida Power, which later found that it was more profitable to generate power than it was to produce ice. It was added to the National Register of Historic Places on November 17, 1982

The Airport

Melbourne was proud of its commercial air service -- Eastern Airlines flew to Atlanta with stops in Daytona, Jacksonville, Waycross and Macon. The waiting room was on the porch of the terminal, with salvaged theater seats. Service was in Martin 202 or 404 piston engine twins and later Convair 340/440's. The hops were so short that the aircraft would leave one airport and barely reach altitude before it was time to dive to the next stop. There were still lots of buildings from the Melbourne Naval Air Station days, including a big hanger with large wooden trusses built of 12 by 12 timbers. New York Giants baseball farm teams were already training there by 1953, because that summer they filmed a movie there called "The Big Leaguer" with Edward G. Robinson, Vera Ellen, and Jeff Rcihards. Some locals had small parts in the movie. The World Premier was held at the Van Croix, complete with a big parade. The Indian River Players took over the old mess hall building and built the stage right over the old steam tables and installed seats rescued from a remodeling of the Van Croix theater.

The Hospital

The hospital in that era was a motel-like structure between US 1 and the railroad tracks north of downtown. In 1954 it had an average of about 25 patients and charged $18.38 per day. It was not unusual for overflow patients to be located on the porch. Construction began in 1961 for a new hospital at the present site on Hickory St. The old structure is still intact off US 1.

Mosquitoes!

Melbourne used to be a part of Mosquito County, and rightly so! Mosquitoes were so bad that in 1943 salt marshes near Cocoa were the first places in the U.S. where spraying with DDT was tried out for mosquito control, and the problem wasn't all that much better by the time of Radiation's start up in the early 1950s. Brevard Mosquito Control had a hangar right across the street from Radiation's Building 2 parking lot, with an elevated tank out back containing the spray used by their spray trucks and biplanes, and nobody objected if you drove up and filled a jug for use around your home. (It could also be used as diesel fuel.) That biplane with pilot Jack Salamela was a welcome sight if you were fishing down at Sebastian Inlet in the late afternoon, and you didn't mind at all when the spray drifted down on you. And kids on bikes would follow the spray trucks through residential neighborhoods. Early mornings, before the wind rose, were prime spraying times, and the sound of the spray plane roaring 100 feet above your roof was a welcome sound, even at 7:00 Sunday morning!

In those days before aerosol cans of OFF insect repellant, the only protection you had was a little glass bottle of oily 612 repellant that you could buy at Harry Goode's. Night fishing at the inlet was a struggle with the mosquitoes. The combination of an onshore breeze, spraying by the biplane, and 612 would do a pretty good job of keeping the bugs at bay until you started back to your car. But then you'd be covered with mosquitoes until you could drive fast for a mile or so with the doors held open to air out the car.

Other Start-ups

There were a few other technical start-up companies in the area. Symetrics Industries started in Satellite Beach and was later located in Indian Harbor Beach and Melbourne.. There is a company of the same name still operating locally, but they have no corporate memory of old times. A Jerry Sinclair (254-0287?) is a recent retiree who may recall some facts. Al Waltz, a Radiator, worked there. Principals included Ed Braun and Art Beach (also an old Radiator).

DBA (Duane Brown Associates) was also active in the early days. I seem to recall them primarily in the optical business. There is a Google finding for a 1962 technical report on the "location and determination of the location of the entrance pupil of the PC-1000 camera" by Duane C. Brown and Ronald G Davis of the Instrument Corporation of Florida, which is also a vaguely familiar name.

Soroban Engineering was formed in 1953 by Charles West from some personnel involved with the design of FLAC (Florida Automatic Computer) at Patrick AFB. (The soroban in the Soroban logo contains the number 1953.) FLAC became operational in 1953. FLAC workers included Duane Brown and Bruce Glass (whose name is also familiar from an area start-up). Soroban modified IBM electric typewriters for computer I/O, as well as designing paper tape and IBM Card punches and readers. Employees at Soroban included John MacNeil (who I still lunch with every Friday), Harland Manweiler and Hasty Miller.

Is anything the same?

Harry Goode's Outdoor Shop is still here, Del's Freeze still looks the same and has the same soft ice cream. Honest Johns still looks much the same, but the entry road has been relocated. There are a number of good old photos in the Melbourne library meeting room. The old Coast Guard station structure between Melbourne Beach and Sebastian Inlet is still standing, and now operates as the Sebastian Beach Inn. The WMMB radio tower probably hasn't changed. And how about Glenn's Tire -- it's been there since the mid-50s.

Chapter 13

Publications

Some of the most interesting artifacts are some company publications of the early years.

Brochures

The oldest company brochure I have is one dating to the mid 1950's, with a stylized guided missile on the cover. It contains a very complete listing of personnel, including the Board of Directors (4 people), the five Company Officers, the six "Administrators," five Consultants, and three Field Representatives. The Engineering Staff runs to five pages. The Products section includes some seldom pictured items, including the flight and lab versions of the Multi-Stylus recorders, the MDC2A Voltage and Current Standard, and the vacuum tube "Force Amplifier" (a strain gauge amplifier), of which I have an example. The brochure includes an impressive listing of past contracts and experience. It also includes what is probably an exhaustive listing of test instruments and equipment owned by the company, three full pages. The brochure concludes with a strong pitch on what a great place central Florida is to live and work.

There are two thick brochures with same cover but slightly different contents, dating to the early 1960's, with a cover photo of the new palm Bay Building 2, and an interior color photo of the Palm Bay complex, with the lake in the foreground. These include an abundance of photos and descriptions of systems delivered. They also include write-ups on the California Divisions, Space Communications and Levinthal. There are a couple of other brochures from this era, often more narrowly focused.

There is also a 1966 document by Larry Gardenhire entitled Questionsa and Answers About Telemetery.

Radiation Ink

There was a lively, informative newsletter, called Radiation Ink, published from 1955 to 1973. I have about 85% of the issues published, and they are a rich trove of information, names and news. Unlike many company documents, Radiation Ink information includes dates, making it particularly useful. Some of the content, like the occasional "Cutie of the Month" is quaintly nostalgic of an earlier and different time.

Magazine Ads

A few Radiation magazine ads have shown up on E-Bay and been purchased by interested individuals. These cite advantages to working at Radiation, and also describe Radiation involvement in important programs from communications and weather satellites to missile systems and antennas..

"Welcome to Radiation Inc."

Brochures of this title were published, evidently as handouts to both visitors and new employees. these are 20+ pages 8.5 by 5.5 stapled publications showing the officers and organization, capabilities summaries (organized by function and physical location), and an area map. These date from the late 1950's.

Technical Papers

Radiation encouraged engineers to publish technical papers at professional conferences. and in peer-reviewed publications. I have a collections of a dozen or so papers, by myself, Jim Proctor, Bill Langford, George Clark, Jim Grey, Bib Cain and Bob Davis.

THE NEW AGE OF COMMUNICATION

Radiation print ads from the 1950's and 1960's

COMMUNICATIONS UNLIMITED
SATELLITES WILL MEAN FEWER BUSY SIGNALS

The decade of the sixties will see both a crisis in communications and its solution. Experts say that by 1962 present transoceanic cables will be saturated by growing message volume. Radio links, too are reaching the limits of their capacity, with little room in the frequency spectrum for expansion.

The answer: Microwave communication by satellite. Microwaves can't be exploited for distant communication in earthbound systems because they travel in a straight line. But every spot on earth will be in point-blank range of satellite relay stations, and thousands of new telephone, teletype and TV channels will be opened to serve man's burgeoning communications needs.

RADIATION Incorporated is at work today on Tiros I and II and Courier, forerunners of the New Age of Communications. The company is recognized by government and missile industry As an important contributor to the present state of the art in telemetry, electronic data processing and antenna systems. If advanced electronics figures in your business, write for our "Capabilities Report." Address Radiation Incorporated, Dept SA-3, Melbourne, Florida.

MAIN OFFICES AND PLANTS ARE LOCATED AT MELBOURNE AND ORLANDO, FLORIDA; AND PALO ALTO, CALIFORNIA.

AT RADIATION, IDEAS BECOME REALITY

Example: Antennas of diversified design

An effective antenna must provide for more than a sum of its parts. For maximum efficiency, it must be considered an integral part of an entire system rather than an individual component. That's why Radiation engineers experienced in complete systems development consistently base final design parameters on ultimate application. The success of this system oriented approach is evidenced by scores of Radiation antennas sweepin the skies from the Pacific to the Atlantic Missile Range.

Thus, Radiation has developed more experience in automatic tracking antenna design and application of telemetry techniques than any other company. A 12-foot dish antenna, featuring new all solid state tracking servo system and mode logic, is now under development for the X-20 (DynaSoar) program. Two high-gain, wideband 85-foot antenna systems will soon be put into service on the Atlantic and Pacific Missile Ranges. Radiation quad-helix array antennas with phase monopulse tracking have been delivered for Bell's Telstar Project. And now, Radiation's broad-band "Telscom" antenna is available. Telscom combine acquisition and tracking operations for telemetry, surveillance and communications and operates from 200 to 2300 Mc.

These examples illustrate Radiation's unique contributions through advanced research and production techniques. If your interest lies in this field, you'll find a challenging and rewarding future at Radiation. Write for more information or send your resume to Personnel Director, Dept SA-4, Radiation Incorporated, Melbourne, Florida. Radiation is an equal opportunity employer.

RADIATION *Melbourne*
A Division of Radiation Incorporated

Ground/Spacecraft Information Handling Systems – Automatic Checkout – RF Systems – Manufacturing

AT RADIATION, CHALLENGE IS OPPORTUNITY
Example: Advanced data systems to speed Minuteman

Minuteman's real plume is a 352-channel trail of telemetry information for instant visual analysis by test crews. The high/low-level PCM multiplexing system — designed and produced by Radiation Incorporated — represents a major advance in data-handling techniques. It is packaged in *less than a cubic foot of space*, processes analog signals, and will yield maximum performance data from each test firing.

Radiation also developed checkout instrumentation to convert Minuteman's 352 telemetry channels into display form for real-time analysis by test crews. And, the company produced four complete ground data-processing facilities to monitor and record information from all digital telemetry and guidance equipment.

Radiation's scientists and engineers have entered the age of satellite instrumentation with competence in data acquisition and processing for aerospace and range instrumentation. Nimbus, Telstar and OAO will utilize Radiation's proven PCM techniques for long life operation as called for in space environments.

Become a part of this challenging space electronics program. We are currently seeking experienced engineers in the design and development of high-speed airborne and ground digital/analog data systems as either individual contributors or project engineers. Send us your resume or write for details. Director of Data Systems, Dept. SA-72, Radiation Incorporated, Melbourne, Fla. *Radiation is an equal opportunity employer.*

RADIATION *INCORPORATED*

Communications systems – Data acquisition and processing – Automatic checkout – RF systems – Manufacturing

AT RADIATION, CHALLENGE IS OPPORTUNITY
Example: PCM telemetry for "Nimbus"

Accurate long-range weather forecasts will be man's best defense against the caprice of the elements. The *Nimbus* meteorological satellite system—being developed by the Goddard Space Flight Center of the National Aeronautics and Space Administration—will improve such forecasts.

Radiation Incorporated was selected by NASA to design and build PCM telemetry for *Nimbus*. The requirements posed a challenge of system long life and high reliability that have led to major advances in the state-of-the-art. For *Nimbus* a new concept of power switching was developed that will result in a power saving of 30:1 over present methods. Other Radiation-built ground systems will process *Nimbus* data.

Nimbus, *Telstar* and OAO (Orbiting Astronomical Observatory) satellite PCM systems are but three of the many exciting projects in which we are presently engaged. If you're the kind of engineer who is stimulated, not stymied, by the myriad challenges of space electronics, you'll find kindred spirits at Radiation Incorporated. If such an environment appeals to you, send your resume or write for information to Personnel Director, Dept. SA-3, Radiation Incorporated, Melbourne, Florida —an equal opportunity employer.

RADIATION *INCORPORATED*

Communications systems—Data acquisition and processing—Automatic checkout—Antenna systems

SCIENTIFIC AMERICAN, March, 1962

THE NEW AGE OF COMMUNICATION

SOON...GLOBAL RANGE FOR THE VOICE OF FREEDOM

In today's race for men's minds, iron and bamboo curtains are not the only obstacles. We are confronted also by an electronic curtain–powerful jamming transmitters that drown the voice of freedom in a cacophony of meaningless sound. And even a powerful medium of information and entertainment in our own country can scarcely cross a border because of limitations to line-of-sight transmission.

Soon three satellites hovering in equatorial orbit above the earth can forever level all these barriers. Functioning as communications relay satellites, they can become the sights and sounds of the free world to the committed and uncommitted everywhere. News, entertainment and such events as the decision of a President can be seen and heard around the world, relayed from space on microwaves impractical to jam. Total cost will be a fraction of that for increasing the capacity of today's overworked communications network.

When? In less than a decade. Communications satellites of the type described here are feasible today. And we at Radiation Incorporated are proud of our part in helping to make them so.

Radiation Incorporated was identified with Score and Courier, each in its way a forerunner of the communications satellites of the 1970's. These and other assignments from government and industry in many areas of sophisticated electronics are a result of our capabilities. If you would like to know about us and our work, write for our brochure, "Capabilities Report." Address Radiation Incorporated, Dept SA-1, Melbourne, Florida.

RADIATION *INCORPORATED*

MAIN OFFICES AND PLANTS ARE LOCATED AT MELBOURNE AND ORLANDO, FLORIDA, MOUNTAIN VIEW AND PALO ALTO, CALIFORNIA

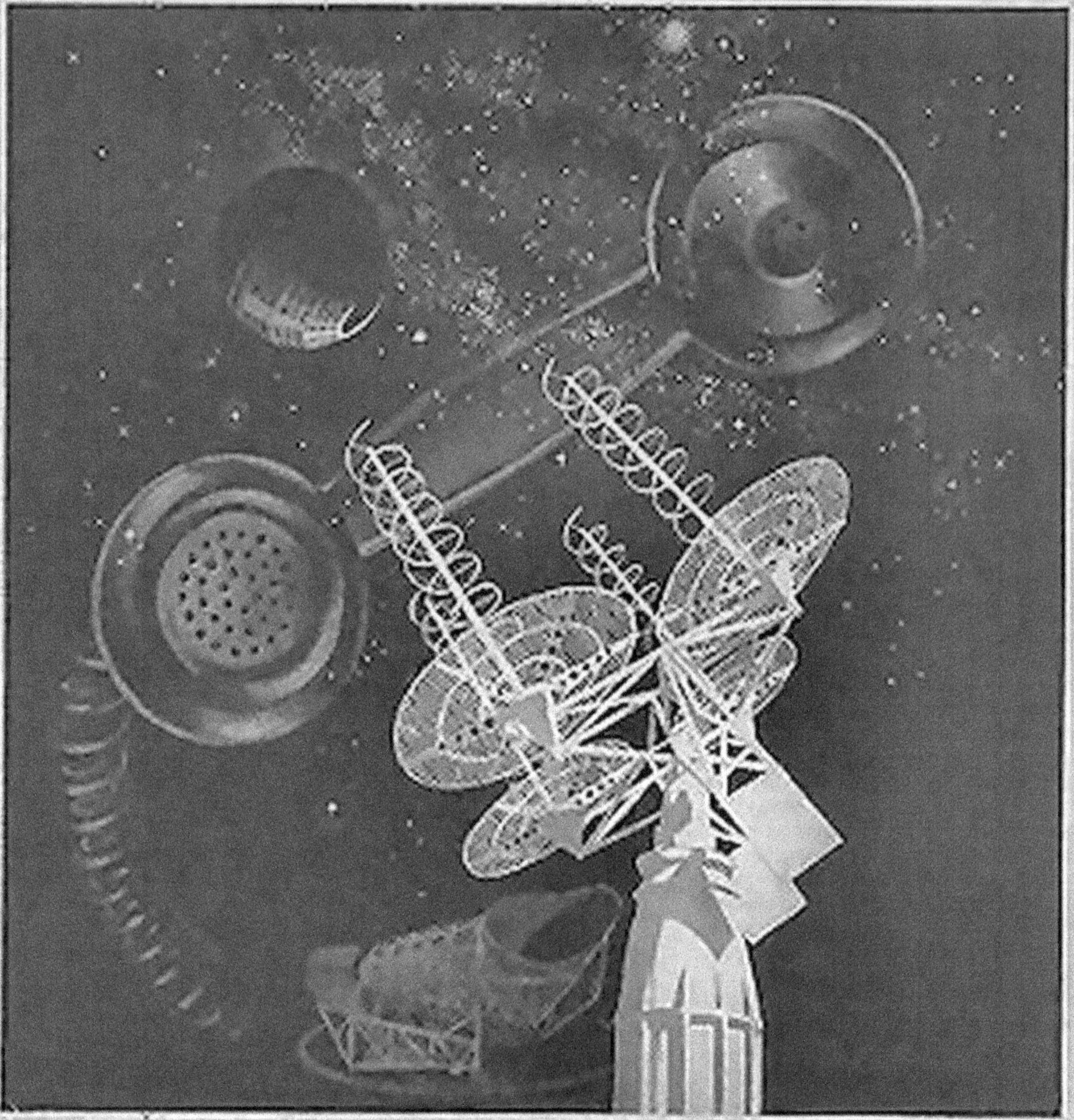

This ad depicts the Quad Helix antenna Radiation built for Bell Labs for use with TelStar, the first commercial communications satellite. The antenna was used to transmit commands to the satellite at 148 MC and receive telemetry signals at 136 MC. It is also used to provide acquisition data to the Bell precision tracker. Radiation also supplied the PCM telemetry system in the satellite.

Ad on a cigarette lighter

Chapter 14

The Ongoing Spirit

Radiation Old Timers Organization

Back in about 1954, as well as any of us can remember, Gantt Hamner invited everyone at Radiation, Inc. (probably fewer than 100 employees at that time) to a "crab-cracking" party at his bachelor house in old Magnolia Manor --- and almost everyone came! People still talk about that party to this day --- having to move the furniture out in the yard to accommodate the crowd, the sight of Algie Douglas hammering the crab claws, Ralph Johnson and Parker Painter lining up the ladies for a good night kiss before they left. There was electricity in the air that night: maybe just smelling (or sampling) those "spirits" that Bill Eddins had distilled on the kitchen stove had something to do with it, or maybe everyone was just in the right mood for a party. Remember the excitement and the enthusiasm of working in a small company where everybody knew everybody---even overtime was fun!

1996 Gathering

Forty-some years later, Gantt, AB Amis, and Mel Cox thought it was about time to get some of the old gang together again for a few hours to rub elbows and swap recollections of "the good old days", so a party was planned at the Amis's place in Grant for Saturday, November 9, 1996. With few exceptions, the invitees were folks who had worked at the "Old Radiation" back in the 50's - before the move down to Palm Bay - and a smiling crowd of just over 200 showed up for an afternoon of fellowship and a seafood stravaganza organized (and largely prepared) by Gantt again. The weather was perfect, the beer was cold, the greetings warm, and all in all it was another magical day.

2000 Gathering

Ty Miller had enjoyed that 1996 get together more than almost anyone, and in late 1999 he'd started talking up another gathering for the early 2000 time frame. But Ty's health was bad, and unfortunately he passed away before he could get it all together. Leif Harris picked up the ball and went forward with Ty's plan, drafting "volunteers" to help him and settling on a plan to have another get together at Gantt Hamner's waterfront place on Palm Bay Point in April 2000. An updated mailing list was developed, building from the list of invitees to the earlier reunion, but physical realities like parking placed a practical limit on the number of people that could be invited, with the result that the crowd at this second gathering was only about half the size of the earlier one -something just over 100 people. Again, the weather cooperated, the setting was beautiful, the catered food good, the beer cold, and the fellowship warm. Ty would have loved it.

2002 Gathering

Participants at the 2000 gathering "voted" to have the next gathering in 2002, preferably at a public facility to spare anyone too much hard work, and capable of accommodating a larger number of people. Recognizing the sad facts of life that our numbers continue to shrink and health issues increasingly limit participation, it was decided to begin an effort to contact and invite everyone who ever worked for RADIATION, INC. while it was still an independent corporate entity. Leif agreed to serve as Program Manager again, and on April 11th, 2002, something just under 200 "old timers" gathered at Lake

Washington County Park on a windy Saturday to share reminiscences and raise a mug to Dave Balser, who had selected and made the arrangements for the park facility, and to other of our friends and heroes who had passed on since our last gathering.

2003 Gathering

Participants at the 2002 gathering "voted" to make these gatherings annual affairs, so Leif again assembled his "Tiger Team" of volunteers and set to work planning and promoting a 2003 get together. Hoping to recapture some of the "warmth" of the original 1996 gathering, the Amises again volunteered to host the gathering at their home in Grant, and efforts were begun to give the event wider publicity and again extend invitations to everyone who had worked at Radiation from inception until the merger with Harris on September 30th, 1967. These efforts paid off on a fine spring day, with something over 200 old Radiators and guests rubbing elbows, sharing reminiscences, and being entertained by a program put together by co-founder George Shaw, who seemed to be relishing his job as Master of Ceremonies.

2004 Gathering

Tiger Team member George Hennessey had made arrangements for the 2004 gathering to be held at the new location, Melbourne Riverview Park, on a brisk spring day, and "almost doctor" Geri Cuthbert did the Master of Ceremonies honors with a program recognizing the unsung heroes of Radiation – the long suffering ladies. But it's getting harder and harder to find folks willing to participate in a "program" – and just as hard to get them to sit still and pay attention to a program anyhow, when they'd really rather be visiting with old friends – so the consensus seems to be to downplay organized programs at future gatherings.

2005 gathering

The 2005 gathering was again held at Riverview Park since this had proved to be an "easy" format, and Radiation employee #24 (hired in 1951) – Ralph Johnson – handled the MC duties with an abbreviated program that allowed more time for "just visiting". (Subsequent get-togethers have also been held at Riverfront Park.)

2006 gathering

We tried for the best of both worlds in 2006 – an entertaining program of live bluegrass/old country music featuring several of our own Radiation Old Timer musicians (backed up by lots of good help) -- but with fast-talking MC Jack Davis keeping things moving right along so there was plenty of time for visiting and still getting home before dark.

2007 Gathering

In 2007, MC Scott Broadway challenged our memories of "glory days" projects, followed by a program of live bluegrass/old country music.

2008 gathering

In 2008 MC Chris Catsimanes humbled us all with his recollection of his family's immigration to the U.S. with only cents in their pocket, followed by bluegrass music off in one corner of the pavilion.

2009 Gathering

In 2009, the MC was Bob Davis and we provided light background music and lots of talk, which worked out better for "just visiting with old friends".

2010 Gathering

Terry Casto MC'ed the 2010 get-together.

2011 gathering

In 2011 Frank Perkins served as host.

This year there were about 125 attendees.

2012 gathering.

The gathering was held on April 21, 2012. Speakers were Jay Fleming, A. B. Amis and George Rassweiller

This gathering concentrates on employees of Radiation, Inc, before the 1967 merger with Harris-Intertype.

2013 gathering

The gathering was held April 20, 2013.with about 118 in attendance. Joe Pira was Master of Ceremonies. It was announced that the group now welcomes anyone employed by the company while the name Radiation was used, i.e. until 1975. Dianne Markham, representing the Brevard Historical Commission was present to make recordings for their Oral History Project. Thanks to all who participated. These recordings is available at the Cocoa library reference departmant.

2014 gathering

The gathering was held on April 17 at Rverview Park. Again, about 118 were in attendance.

The get-togethers are organized by a "Tiger Team" of volunteers.

Photos and videos have been taken at the gatherings over the years, and represent a valuable record of people over the years.

Chapter 15

Musings

In the course of reviewing and remembering Radiation history, a number of factors came to mind which I believe are significant to to the success of a company or other organization. I would like to offer my musings on these topics in the hope they might be applicable in a broad sense. The lessons to be learned from a company's history are best understood after the consequences have played out and emotions have cooled.

The first observation relates to how jobs or programs were managed. In the early days, the Project Engineer was responsible for everything about a program, including staffing, planning and budgeting and everything about the design. This is not to say that there were not specialists in these supplementary areas, but they worked under the direction of the Project Engineer.

Later, it became the fashion to have Program Managers run jobs, with design and engineering mere cogs in the organization. Then management schools proclaimed the self-serving opinion that management was the key, and that a "good manager could run anything." This led to situations where Program Managers led jobs they were incapable of understanding, much less contributing to the design, even when the engineering design was the principal challenge of the job.

I don't mean to belittle managers in general, but to point out that the concept that "a good manager can mange anything" is a dangerous concept for a company, or a government for that matter. In many complex situations, an understanding of the technical issues and problems is necessary to devise a good solution.

An outgrowth of this management emphasis was the "matrix" organization, whereby management and engineering were separated in the corporate structure, with the engineering organization treated as a holding area for engineers, who would be assigned to a program manager for the duration of a job, and then returned to the pool. Intended to improve utilization of engineering talent, this scheme actually killed the entrepreneurial spirit and motivation of the engineers. There was no motivation for engineers to work on follow-on aspects of a job when they would be re-assigned to something else. Customers craved the continuity of the engineering effort and would go out of their way to award follow-on jobs to a proven engineering team. This effect was lost when they realized that the company intended to staff the follow-on with a different complement of engineers.

A related issue has to do with the development and handling of individuals in an organization. There is a tendency for an organization to categorize individuals, for example, as a manager or a direct contributor, or a systems engineer or a logic designer. Clearly, different individuals have different skills, but mis-identifying the skill, or trying too hard to bend the skill to the priorities of the organization can lead to trouble. A good engineer may have talents that can be used in marketing, but forcing him out of engineering into marketing is a mistake if he strongly prefers to be an engineer. Likewise, ignoring the ability of an engineer to develop rapport with customers (because he isn't in marketing) is counter-productive. Often a little effort can obtain useful results from an individual the "system" has labeled marginal. I personally recall getting very useful and badly needed checkout work from an engineer rejected by most other leaders in the organization, because I chose the tasks to suit his personality and capabilities.

One of the factors in Radiation's success was an effective profit sharing plan, the Employees, Retirement and Benefit Plan (ERBP). It was started in 1952. The company contributed up to 25% of profits to a trust fund, credited to eligible employees in proportion tho their salary. . Employees could also contribute up to 10% of their earnings, on a pre-tax basis, to the fund. The assets of the fund were managed by Trustees appointed by the Radiation board of directors and often achieved good returns on stocks, real estate and leasing companies.

The ERBP was a powerful incentive for employees to maximize company profits by their actions and performance. The plan also served as an attractive incentive for scarce technical talent to come to work for Radiation. Retirement plans are common today, but the structure of the Radiation plan seemed to resonate well with employees. Its features and benefits were well-publicized in company mailings and newsletters.

An October 1963 publication read:

FUND CONTRIBUTION REACHES RECORD HIGH
The Company's contribution to Radiation Employees Trust Fund reached $449, 929 for fiscal 1963, largest dollar amount ever received by the Fund.
This figure represents 5.4 percent of participating wages which totaled more than $8.3 million.
Also announced was the year-end unit value which jumped seven cents to $6.37 as of Aug. 30. The unit value increased 27 cents or 4.4 percent during the 1963 fiscal year.
Previous Company contribution high of $306, 000 came in 1961.
Also, the record contribution figure is just $71 dollars short of $450, 000 mark.
Roy Bixby, Trust Fund trustee, said the reason for the increase in the unit value largely resulted from a rise in the stock market. It was just a year ago that the same market dropped considerably to force down the unit value.
Bixby also said a Trust Fund annual report will be available sometime in November and is expected to review the Fund's 1963 activities and list the stock holdings.
Annual statements to all participants showing the number of new units and the cumulative unit total will be going out soon, Bixby said.
The Fund announcement comes on the heels of a year-end earnings report by Homer Denius, Radiation president.
Announced last week, Radiation earnings reached $1, 072, 287 or $1.03 per share on sales of $30.6 million.
Included in the $1.03 per share earnings was 30 cents per share from the sale of Radiation at Stanford.
The 1963 earnings tops last year's figure by almost a half million dollars. In 1962, Radiation earnings of $271,914 or 27 cents a share came on sales of $24,177,045.
Denius said consolidation of activities in Melbourne, start of Radiation's microelectronics operations and a major growth in production were three changes that led to increased efficiency and sales.

Cultivating and weeding divisions
As a company grows, there is a tendency and need to form new divisions. However, a division, even a successful one, can sometimes threaten the well being of the overall company. One of the early Radiation remote divisions was the Orlando Instrumentation Division. This was formed to perform on a contract for instrumenting jet aircraft, which could not be handled by the runways in Melbourne at the time. As time went on, the work being done in Orlando became more and more similar to work being done in Melbourne. This drift toward similarity (and competitiveness) is a common occurrence, probably due to the overall company interest, capability and reputation in these common fields. Separate organizations working on the same problems, and familiar with each others work tend to develop strong opinions that their way is better. This led to Orlando and Melbourne having very similar, but not identical logic cards. This prevented the common production of logic cards in larger and more efficient quantities.
Much later, in Harris days, a similar inter-divisional conflict treaded on even more dangerous ground. Harris had acquired RF Communications Division, in Rochester, NY. One of their areas of expertise was HF radio, the frequency rang from 3 to 30 MHz. [As an aside, the band was so named when it really was high frequency. Technology advanced and higher frequencies became possible and useful, leading to a proliferation of modifiers to the band names: Very High Frequency (VHF), Ultra High Frequency (UHF), Super High Frequency (SHF) and on and on. Names die hard even when their rationale ceases.]
Melbourne had never done much work in HF, so there wasn't significant conflict. However, a technology path in Melbourne in a non-related field led, through customer queries and encouragement, to a breakthrough technology in a modem for the HF channel (which has unique characteristics).

Meanwhile, Rochester had also been working on advanced HF modem technology and arrived at an entirely different solution. Both of the Harris modems were vastly superior to any external competition. However, the competition between divisions led to marketeers from both division belittling the oppositions modem, implying some hidden flaws. Such negative marketing is seldom effective, but is poison when applied between divisions. Doubts planted in a customer's mind are applied to the entire company, with the result that they go to someone else entirely--certainly not the best result for Harris.

The moral to all of this is that upper management needs to monitor divisions closely (with the big picture in mind), and quickly rectify inefficiencies and counterproductive tendencies. Not easy, but necessary.

Divisions are sometime created for geographical diversity, or to get closer to a customer complex, or to appeal to a different employee pool. These can be valid strategies , but the consequences must be faced. The cultural differences can be significant: people and organizations are different between Florida, California and New York. This, combined with distance makes the likelihood of misunderstandings larger.

A final muse has to do with the "merger" of Radiation and Harris-Intertype. The term was inaccurate, as Radiation became a division of Harris-Intertype, but was apparently used to mollify advocates of an independent Radiation, of which there were many, including founder George Shaw. Homer and others apparently felt that internal growth would not be sufficient for Radiation to finance necessary technical advancements, primarily in integrated circuits, or micro-electronics as it was called then. It was felt that an internal microelectronics capability would be necessary to succeed in future aerospace systems jobs which were Radiation's bread and butter, and that achieving this capability would be very expensive. It turned out that commercial and consumer demand drove and funded integrated circuit development and no aerospace company needed or achieved integrated circuit manufacturing expertise.

On the other hand, Harris-Intertype viewed the acquisition as necessary to add electronic capability to the existing printing press capability to form a broad-ranging communications systems company. It turned out that the mechanical aspect of the business withered away, for various reasons and Harris became more like Radiation than vice versa, to the point of having a headquarters building on Melbourne airport property near the original Radiation plants.

Chapter 16

Where to Find More Material

This chapter contains a general description of known locations of Radiation historical material.

The most readily accessible material is on line at WWW.Radwiki.Wikispaces.com

The most complete hard copy files are currently in the care of the author, Frank Perkins. This consists of about 10 linear feet of letter size files, composed of loose documents, bound reports and other documents and photographs. Most of this material has been sorted by catagory. The curent plan for a permanent home for this material is the archives of the Brevard Historical Commission at the Central Brevard Library, 308 Forrest Ave., Cocoa, FL 32922 (phone 321-633-1794)

He also has a a collection of CD's and DVD's of Old Timer meetings, as well as a few items of hardware, including:

A display case of Radiation PC cards and modules from the 1950's and 1960's.

A vacuum tube "Force Amplifier" and associated meter type display. This is actually a strain gauge amplifier used in stick and rudder force systems built in the 1950's.

Several front panels of Radiation equipment and products.

Several other people have personal collections:
.
a. Nancye Meyers and Tim Hartsfield at Harris Corp in Palm Bay, FL have some documents including Radiation, Inc. Annual Reports, some copies of Radiation Ink, the company newsletter, old advertisements, miscellaneous documents, some Radiation-built hardware from the 1950's.

b. Jamie Cox (321-255-5387) (son of Mel Cox, an early Radiation employee) has an indexed collection of documents and some examples of Radiation-built hardware from the Vietnam era. He also has a number of Radiation history photos on https://www.flickr.com/gp/ajmexico/0ZLu74.

c. Jerrie Hixon (Melbourne), daughter of founder George Shaw, has a collection of George Shaw's personal documents.

d. Katie Nunez, daughter of Hans Scharla-Nielsen, a very early Radiation employee, has a collection of Hans' documents.

e. Steve Cook at Harris (321-729-2232) has a collection of of hardware that Radiation built for the Apollo program.

f. The Evans Library at Flrida Institue of Technology in Melbourne has a couple of file drawers of of partially sorted material from Frank Perkins collection..

g. Old Radiator George Rassweiler (321-723-8364) has a collection of DVD's and audio recordings of Old Timer meetings.

Material is available n line at WWW.Radiwiki.Wikispaces.com

www.ingramcontent.com/pod-product-compliance
Lightning Source LLC
Chambersburg PA
CBHW081110170526
45165CB00008B/2403